NEW PERSPECTIVES ON GALILEO

THE UNIVERSITY OF WESTERN ONTARIO SERIES IN PHILOSOPHY OF SCIENCE

A SERIES OF BOOKS

ON PHILOSOPHY OF SCIENCE, METHODOLOGY,

AND EPISTEMOLOGY

PUBLISHED IN CONNECTION WITH

THE UNIVERSITY OF WESTERN ONTARIO

PHILOSOPHY OF SCIENCE PROGRAMME

VOLUME 14

NEW PERSPECTIVES ON GALILEO

Papers Deriving from and Related to a Workshop on Galileo held at Virginia Polytechnic Institute and State University, 1975

Edited by

ROBERT E. BUTTS

Department of Philosophy, the University of Western Ontario, London, Canada

and

JOSEPH C. PITT

Department of Philosophy and Religion, Virginia Polytechnic Institute and State University, Blacksburg, Virginia, U.S.A.

D. REIDEL PUBLISHING COMPANY

DORDRECHT : HOLLAND / BOSTON : U.S.A.

Library of Congress Cataloging in Publication Data

Main entry under title:

New perspectives on Galileo.

(The University of Western Ontario series in philosophy of science ; v. 14)
Bibliography: p.
Includes index.
1. Science–Methodology–Congresses.
2. Science–Philosophy–Congresses. 3. Galilei, Galileo, 1564–1642. I. Butts, Robert E. II. Pitt, Joseph C.
Q174.N48 509'.2'4 77–17851
ISBN 90–277–0859–2
ISBN 90–277–0891–6 (pbk.)

Published by D. Reidel Publishing Company,
P.O. Box 17, Dordrecht, Holland

Sold and distributed in the U.S.A., Canada, and Mexico
by D. Reidel Publishing Company, Inc.
Lincoln Building, 160 Old Derby Street, Hingham,
Mass. 02043, U.S.A.

Printed in The Netherlands

TABLE OF CONTENTS

PREFACE	vii
ROBERT E. BUTTS and JOSEPH C. PITT/Introduction	ix
WINIFRED LOVELL WISAN/Galileo's Scientific Method: a Reexamination	1
ROBERT E. BUTTS/Some Tactics in Galileo's Propaganda for the Mathematization of Scientific Experience	59
WILLIAM A. WALLACE/Galileo Galilei and the *Doctores Parisienses*	87
WILLIAM R. SHEA/Descartes as Critic of Galileo	139
PETER MACHAMER/Galileo and the Causes	161
JOSEPH C. PITT/Galileo: Causation and the Use of Geometry	181
H. E. LE GRAND/Galileo's Matter Theory	197
ERNAN McMULLIN/The Conception of Science in Galileo's Work	209
INDEX	259

PREFACE

The essays in this volume (except for the contribution of Dr. Le Grand) are extremely revised versions of papers originally delivered at a workshop on Galileo held in Blacksburg, Virginia in October, 1975. The meeting was organized by Professor Joseph Pitt and sponsored by the Department of Philosophy and Religion, The College of Arts and Sciences, and the Division of Research of Virginia Polytechnic Institute and State University. The papers that follow deal with problems of Galileo's philosophy of science, specific and general problems connected with his methodology, and with historical and conceptual questions concerning the relationship of his work to that of contemporaries and both earlier and later scientists.

New perspectives take many forms. In this book the 'newness' has, for the most part, two forms. First, in the papers by Wisan, Shea, Le Grand and Wallace (the concerns will also appear in some of the other contributions), greatly enriched historical discoveries of how Galileo's science and its methodology developed are provided. It should be stressed that these papers are attempts to recapture a deep sense of *the kind of science* Galileo was creating. Other papers in the volume, for example, those by McMullin, Machamer, Butts and Pitt, underscore the importance of this historical venture by discussing various aspects of the philosophical background of Galileo's thought. The historical and philosophical evaluations and analyses compliment one another. We think it important that they should – Galileo's thought was much too complex, much too original *and* much too dependent upon earlier modes of enquiry to make the task of reconstructing his philosophy of science easy. No catalogue of historical details will do that task; nor will a systematic presentation by philosophers of the 'logic' of his methodology. The authors of the following essays remain faithful to the ambiguity that was characteristic of Galileo's work as a productive scientist. In this way they manage to instruct us best in the many dimensions of one of the seminal persons involved in creating modern science. All too frequently philosophers and historians are inclined to go their separate ways. It is our hope that the present volume will be seen as a good example of what can be done in the difficult job of building a crossroad.

This volume would not have been possible without the extraordinary cooperation of the authors of the papers, many of whom commented on and criticized portions of the papers of others involved in the workshop. The workshop itself was only possible because of the considerable support offered by Virginia Polytechnic Institute and State University through the good offices of Associate Dean of Research for Arts and Sciences Thomas Gilmer, Dean of the Research Division Randal M. Robertson, and Dean of the College of Arts and Sciences William C. Havard. We thank them for that support.

May, 1977

THE EDITORS

INTRODUCTION

The now classical works of Koyré and Crombie, works different in philosophical and historiographical approaches, provoked new scholarly interest in Galileo's science. The motivations of their work were, however, very similar. Both wanted a more detailed knowledge of Galileo's actual achievements in positive science, both sought to understand the contribution to scientific methodology generated by this major figure participating in 'the scientific revolution,' and both, by means of novel insights and conceptual combinations of historical and philosophical materials in ways undreamed of by earlier scholars, introduced a whole new dimension to thought about science that can be called the 'metaphysics of science.' Students of the scientific revolution and the work of Galileo fully realize the importance of the contributions made by these two men. The point is that no one could have forecast the impact that their work – and that of others – would have on our attempt to understand the beginnings of modern science. At the present writing, there is probably no issue that receives greater attention in thought about science than the attempt to understand scientific change, scientific progress, and like concepts. That attention, moreover, is beginning completely to overshadow interests that were once dominant, for example, problems of confirmation in science, the status of hypothetico-deductive explanation, or the 'meaning' of hypothetical constructs having to do with the empirically unobservable.

We can of course mention others whose keen historical or philosophical sense has led us to think new thoughts and envision new possibilities: Kuhn and Burtt are only two. But this 'Introduction' is not the place to rehearse, even in a summary, the major contributions that have been and are now changing the character of history of science and philosophy of science. However, it does suit the purposes of this volume to indicate that the present scholarly forms of interest in the science of Galileo exist in an enriched context of a fairly widespread exploration of just those dimensions of Galileo's career exposed in an exciting way by Koyré and Crombie. In this connection it does seem appropriate to name some names (with apologies to those neglected for reasons of space or because of simple oversight). There are three groups of contributors to mention.

No one will deny that many of the most important studies of Galileo's positive scientific discoveries have been made by Stillman Drake. Equally important work has been done by William Shea, E. J. Aiton, A. Rupert Hall, Thomas B. Settle, and others. For many of these scholars the question of *why* Galileo accepted the conclusions he did transforms a reporting of what he actually says he did into an attempt to comprehend the methodology he is reputedly so famous for having developed. One important aspect of this methodological concern has to do with experiments and what Galileo thought they accomplished. It is one thing to ask whether or not he dropped the steel ball from the Tower of Pisa; it is quite another to ask why he would have thought it important to do so just in case he ever did. Drake has made meticulous studies of the materials which appear to show that Galileo actually did perform experiments, others, for example, Settle, have tried to reproduce what they think must have been Galileo's experiments.

Galileo's 'experimental method' is obviously an important feature of his science. Important as this method is, however, discussions of what experiments he attempted and how he apparently construed the reputed results, do not rule out what are probably deeper considerations of method – considerations we are quite prepared to call 'philosophical.' Just as questions of what actually was discovered blend into questions about what the discoveries mean, so questions about what the discoveries might mean blend into questions about how the discoverer conceived the universe in the very construction of his questions. This very complex blend of methodological and 'metaphysical' questions constitutes the third dimension of contemporary interest in the accomplishments of Galileo. Indeed, it would be our contention that more important work has been offered in this area than in any other. We have mentioned the classical 20th-century thinkers who have contributed memoirs on the frontier of methodology-metaphysics. Others of major importance are Winifrid Wisan, William Wallace, William Shea, and a philosopher who keeps coming back to haunt us with his challenge that science, methodology and metaphysics are all bright meadows that will turn into swamps, all hopes that will become utter disappointments – Paul Feyerabend.

We have suggested that three areas of Galileo's work have come into prominent focus during this century: (1) the novelty of his positive scientific discoveries (if any); (2) the nature of his way of doing science – his methodology; (3) his 'metaphysics'; the conditions, if you will, under which he would have found any science acceptable. The three categories must of course be seen in the larger context of what others were attempting in science during his lifetime, and in the framework of the kind of revolution in thought about nature that we now take as an obvious feature of his concerns. We shall turn away from these more general considerations in order to show respects in which the contributions to the present volume continue the vigorous activity that has been going on in Galileo scholarship during the last few decades.

For the most part the papers in this volume deal broadly with the philosophical background of Galileo's thought and with questions concerning Galileo's methodology. Two papers, those of Shea and Wallace, appear to be more or less standard historical studies, Wallace concerned with dating Galileo's so-called *Juvenalia*, Shea investigating the question whether Galileo and Descartes independently both discovered the times-squared law. In this case appearance is not reality; the two papers also show implications for understanding Galileo's methodology and metaphysics. It is striking that the array of papers exhibits a coincidence of similar insights and interpretations and an overlap of attention to common themes. We propose to look briefly at three such themes: Galileo's method, his views on causality, and his treatment of the nature of mathematics. As we shall see, the three themes are themselves interrelated and cannot neatly be compartmentalized.

In her perceptive study of the development of Galileo's method over a 50-year period, Wisan outlines the basic points. The model for Galileo's methodology is mathematics, not logic. Unlike most of his contemporaries operating in a quasi-Aristotelian methodological tradition, he thought that knowledge came through the application of mathematics to experience – knowledge is the result of calculation or observation of experimental results, it is not the grasping of essences. (The profound association between this form of essentialism and the subject-predicate logic of Aristotle is attested to by the large variety of metaphysical systems based on this logic leading from Aristotle to Bradley. That the logic is already to be found in Plato is a historical commonplace.) Wisan goes on to spell out in detail how Galileo wanted rational principles to be rendered immediately evident through reason and experience, and how discovering true causes means discovering which known rational principles provide grounds for the events to be explained.

Shea shows that there was little contact between the younger Descartes and the work of Galileo. One wonders why? If Wisan is right and the point of Galileo's science was to render results immediately evident, is this not much like the Cartesian methodological insight that mathematical knowledge is gained through intuition? Leaving our readers to ponder this speculation, we return to Wisan's account. It follows from her account of the desiderata of Galileo's method that that method is not comparable to later formulations of what we now call the hypothetico-deductive model. Rather, the method of making basic principles evident derives from Galileo's conviction that in mathematics we can proceed to evident conclusions on the basis of moving through small and necessary steps from defined terms and explicit assumptions. Her novel conclusion is that Galileo in effect transfers this kind of mathematical model to physical science, so that inferring causes from observed effects takes the same structural form.

Wisan's characterization of Galileo's methodological motives is confirmed, although obviously for different, even sometimes conflicting, reasons by a number of other authors of papers in this work. In the context of his elaborate historico-philosophical argument that Galileo's main intellectual impulses are derived eclectically from scholastic Aristotelianism and not from the 14th-century *Doctores Parisienses* (a conclusion of not inconsiderable philosophical importance in its own right), Wallace reintroduces a point for which he has argued on earlier occasions, namely that for Galileo reasoning *ex suppositione* (a form of reasoning classically regarded as being only tentative logically) is demonstrative. (One should note that McMullin disputes this conclusion, arguing at some length and with persuasive logic that for Galileo *ex suppositione* reasoning was, as it was for many of his contemporaries, not demonstrative, but hypothetical.) Machamer suggests that for Galileo what Aristotle called 'knowledge of the reasoned fact' is equivalent to mathematical demonstration. Shea points out that Descartes' failure to find a way to manage the times-squared law mathematically contrasts with Galileo's mathematically more rigorous treatment. Butts, in his detailed analysis of Galileo's argument for the applicability of mathematics to the world, presents a good example of the point made by both Wisan and McMullin that for Galileo the job is to render basic principles immediately evident.

For the most part, the authors included in this volume agree with Wisan's reconstruction of Galileo's methodology. There are of course differences in emphasis. For example, both McMullin and Wallace lay more stress on the Aristotelian background of Galileo's methodological style. Everyone agrees that Galileo's greatest strength lay in a gifted application of geometry to physical problems. But the nature of that application – indeed, the nature of that geometry – is disputed. Of great importance is the rich study by McMullin of what he calls Galileo's 'two conceptions of science.' On McMullin's reading Galileo's methodological career was dominated by the Greek ideal of a demonstrative science. However, faced with explaining phenomena that are physically remote, invisible, or enigmatic (the motions of the earth), Galileo introduced a second method similar to Peirce's 'retroduction.' A much enlarged understanding of Galileo's science emerges from a close comparative study of the papers by Wisan and McMullin.

For many years textbook accounts of Galileo's philosophy of science attributed to him the expurgation of final causes from the book of nature: from now on science will deal only with efficient causes, nature will no longer be understood teleologically. From whatever source this old chestnut comes, and whatever are its apparent merits, it contains some philosophical worms. If it is true that Galileo's model for both methodology and the nature of reality is mathematics, is it not clear that the central focus of his thought about causes would have been on *formal* ones? When philosophical concentration upon efficient causes becomes prominent in the Seventeenth Century, it is because physicists attempt to reduce all forms of physical phenomena to cases of impact or collision. It is not at all clear that Galileo adopted this view of a privileged physical model, especially since – as LeGrand points out in his paper – for Galileo the material atoms are transposed into atoms regarded as *infinitely* small, that is, as *mathematical* points. On such a theory it would be difficult to endorse collision as the basic explanatory model, and hence difficult to champion efficient causality as philosophically central. However one deals with these (apparent) difficulties, one will also have to deal with Stillman Drake's attempt to dispel the entire force of the textbook account in his argument that Galileo was not concerned with any form of causality at all.

Three of the papers in this volume address themselves directly to the question of Galileo's philosophical attitude to causes. Machamer, citing earlier work by Wallace, presents evidence that Galileo was indeed interested in causal explanation in science. Machamer's interesting point is that given Galileo's identification of mathematics and experience, the important causes are the formal ones: efficient causality plays no important role at all in this scenario. McMullin gives general support to Machamer's contention that Galileo did not abandon causal explanation. He provides an elaborate analysis of Galileo's use of retroductive causal explanation, claiming that this method is one of the two prominent aspects of Galileo's approach to science. Pitt suggests that Galileo's revolutionary change in ways of doing science involved the abandonment of final causes, replacing explanations in such terms with 'mathematical descriptions' of the phenomena. There is much to consider with regard to questions about the role causality plays in scientific explanation. Pitt suggests that perhaps the key ingredient in a scientific revolution – a massive change from one way of doing science to an altogether different one – is a change in ways of thinking about what causes are (or ought to be regarded as being).

If Wisan and Shea are right about Galileo's mathematically modelled methodology, and if Butts is correct in his portrayal of Galileo as being willing to reject the world – rather than mathematics – if the world tells us something that is false in mathematics, then surely one can make out a good case for the positions of both Machamer and Pitt: for Galileo, thought about causality can no longer be captured in Aristotelian terms. Mathematical physics – a science which for Aristotle and most medieval Aristotelians had a merely descriptive status with no ontological force – becomes the paradigm science for Galileo; and with the establishment of this paradigm final causes become irrelevant along with efficient causes. Indeed, if one enjoys precedent hunting, he might want to say that Galileo, along with Bertrand Russell, concluded that causal talk is really eliminable from mature science. (Galileo had not advanced to this point; taking into account Wallace's point that Galileo remained throughout his career in the grips of a mixed Aristotelian genetic inheritance, the different but reinforcing approaches to causality in Galileo in the papers by Machamer, McMullin, and Pitt reveal metaphysical postures not to be neglected in attempts to understand Galileo's metaphysical 'location' in early modern science.)

Much of what has already been said about the papers in the volume points to a very generally accepted view of the science of Galileo: he was one of the first major scientists to insist that science is thoroughly mathematical. Mathematics must be used in the determination of experimental limits, it must be used in calculating the implications of experimental results, and it must be used for the very specification of what counts as an 'object of reality.' Nearly all of the present collection of essays has something to say to this point. The essay by Butts concentrates on this issue, attempting to show in detail how Galileo argued for the applicability of mathematics to the observed world (given that the world does not contain any objects like circles, spheres, and other 'mathematical objects'), and also made the more contentious claim that reality – the reality knowable by man – is essentially mathematical. Shea reinforces this interpretation by contrasting what he calls Galileo's 'mathematical platonism' with Descartes' failure to commit himself to mathematical realism. We have already discussed Wisan's contention that Galileo's entire methodological development involved an essential recognition of the centrality of mathematical methods. Pitt claims that acceptance of mathematical descriptions of experimental results in the science of Galileo (descriptions given in the geometry of his day) eliminated altogether any appeal to teleological considerations. Machamer rejects the problem Butts worries about – the applicability of mathematics to the world – resting content with Galileo's identification of mathematics and experience.

Also impressive is the position of LeGrand, whose argument is entirely devoted to defending the thesis that Galileo introduced a novel theory of matter. In attempting to show that Galileo's confidence in mathematics is partly expressed in his acceptance of one form of the primary-secondary quality distinction, Butts puts emphasis upon what are perhaps incidental passages in *The Assayer*. LeGrand, moving from more substantial passages in Galileo's mature work, argues that he was not an atomist after the fashion of Democritus (whose position does seem to depend crucially on the distinction between types of qualities), but that his theory of matter is unique in suggesting a new metaphysical construal of atoms as infinitely small, that is, as mathematical points. We can only comment that the distinction between primary and secondary qualities (or whatever you want to call them) in Galileo shades off into insignifigance when the total ensemble of his thoughts about the primacy of mathematics is seriously entertained.

If matter is decomposible into mathematical points, if both mathematical theory and mathematically designed experiments are to be preferred methods of science, and if the Book of Nature is written in mathematical terms, then there can be no doubt that one of the leading tenets of Galileo's philosophy of science proclaims both the methodological and metaphysical advantage of regarding 'saving the appearances' as 'confirming the realities.'

During the period when he was regarded as the 'father of modern science' Galileo was a much neglected figure in the history of science, a fate not unfamiliar to fathers of anything at all. It is a good thing that Galileo can now be thought of as someone who did a bit of good science in his day, and that the way in which he did it was sufficiently different to make it important. Contributors to this volume do not polish Galildo's tombstone, they seek to understand what he was trying to do. Perhaps the best compliment one can pay to a departed father is to laugh at him; the laughter of the good historian and the good philosopher is to be heard in the vibrations of the inquiry: Why did he think that? What motivated him to try that? Where was he when? Who did what to him where? Could he really have believed that? and on and on. Kant was right: understanding unique properties of an individual is an art, not a science. The papers which follow are exercises of that art. We think they are important instantiations of that art.

ROBERT E. BUTTS
JOSEPH C. PITT

WINIFRED LOVELL WISAN

GALILEO'S SCIENTIFIC METHOD: A REEXAMINATION*

INTRODUCTION

There is general agreement that Galileo is one of the most important figures in the history of science and scientific method. His science, method and thought seem clearly to reflect critical aspects of the transition from medieval to modern science. But here agreement ends. There is no consensus concerning the exact nature of the changes which took place or Galileo's precise role with respect to these changes. And what Galileo thought on almost every topic is subject to endless controversy.

One of the main difficulties lies perhaps in Galileo's failure to provide a systematic account of his views. Brief comments on philosophy and method are scattered throughout his writings and occasionally one finds a paragraph or so of sustained discussion. But these passages are often obscure, usually fragmentary, and sometimes contradict one another. Consequently they have provided the basis for radically different interpretations covering virtually the entire spectrum of philosophical stances and methodological precepts. As A. C. Crombie points out, philosophers "looking for historical precedent for some interpretation or reform of science which they themselves are advocating, have all, however much they have differed from each other, been able to find in Galileo their heart's desire" (1956, p. 1090). Crombie might well have added that historians have done much the same thing, for even if they have no philosophical axe to grind, nonetheless they see Galileo through eyes accustomed to some particular interpretation of the scientific enterprise.[1]

Moreover, to the historian Galileo presents special difficulties. His writings are voluminous and extend over a period of some fifty years, reflecting many changes in his thought and exhibiting both real and apparent contradictions on crucial issues. Our difficulties in understanding this material are compounded by insufficient knowledge of Galileo's immediate background. To reconstruct his work and thought we need to know what he was trying to do and what he did, and our main clues, of course, are from what he wrote. But

R. E. Butts and J. C. Pitt (eds.), New Perspectives on Galileo, 1–57.

to decipher his writings we must know what the language he used meant in his own time. And this requires more detailed and comprehensive knowledge of Galileo's immediate intellectual background than we yet have. To understand this man – one of the last of the Renaissance figures who were deeply immersed in most aspects of the culture of their time – one must know a good deal about the philosophy, theology, literature, arts and crafts, as well as the mathematics, science and engineering current in his day. Nor can one neglect the social and institutional background of the period.

Much work has, of course, been done in all these areas and many scholars today are engaged in very promising investigations.[2] Background material alone, however, no matter how great in quantity or quality, will reveal little without a precise knowledge of what Galileo himself said and did. Without accurate citations from Galileo's own writings, placed in correct context, examination of those whom he read, listened to or talked with will prove fruitless. And much still remains to be done in correctly analyzing the content of Galileo's texts. A great deal is known, for example, about the resolutive and compositive methods, both in medieval and early modern writings, but Galileo's use of these terms has been greatly exaggerated and the context misunderstood, thus giving rise to distorted interpretations of his methodological remarks (see Sections I and IV of this paper). Similar difficulties infect more recent interpretations of Galileo's philosophical attitudes and scientific procedures.

It seems clear then that in addition to further background studies and 'external' histories of Galileo's period, we still need deeper and more comprehensive 'internal' examination of Galileo's writings. In my study of Galileo's science of motion I attempted to fill some gaps in our knowledge of Galileo's work and provide a more secure base for further understanding his use of mathematics in the development of mechanics. Although it is not possible to draw sweeping conclusions from study of Galileo's writings on motion alone, detailed examination of these writings suggest that both his aims and his methods were quite different from those generally supposed.

In this paper I will attempt to enlarge the picture by sketching the main outline of Galileo's methodological development over a fifty year period from his earliest writings on motion and cosmography in Pisa and Padua (see Section I), his early work in Florence on hydrostatics (Section II) and astronomy (Section III), to his great *Dialogue* (Section IV) and the *Two New Sciences*

(Section V). The results will, of course, remain incomplete pending further insights which may be drawn from more extensive studies of Galileo's own writings as well as from closer examination of the writings of his most likely predecessors and other influential figures. However, special attention will be given to those aspects of Galileo's thought which have given rise to the most serious misunderstandings and misinterpretations, and some tentative conclusions will be drawn in the hope that they may be of value in guiding further inquiry into his immediate background.

One of the chief sources of difficulty in interpreting Galileo stems from the choice of materials on which various analyses are based. Galileo's writings deal mainly with astronomy and mechanics and for the most part he employed different methods and applied different standards for work in these two fields. In astronomy, for example, he mixed sound scientific reasoning with dubious arguments which he hoped would persuade his readers of the truth of the Copernican system. In mechanics, on the other hand, he consistently sought to produce a rigorous mathematical treatise. Here is a source of much misunderstanding. Most discussions of Galileo's science and method are based primarily on his work in astronomy, especially his best known work, the *Dialogue* on the Copernican and Ptolemaic systems, whereas relatively little attention is given to the treatise on motion in the *Two New Sciences.* Various passages in the latter work have, of course, been closely examined, but for fuller understanding of this side of Galileo's scientific activity it is essential to look at the 'new science' as an integral whole and to see how Galileo gradually put together a treatise which he believed comparable to the mathematical works of Archimedes.

Also, while much has been said concerning Galileo and mathematics (and Archimedes), it has been insufficiently noticed that to an important degree his 'mathematicism' consists in the attempt to reduce natural science to the Greek mathematical model in order to achieve the logical certainty of mathematics. To do this meant substituting the kind of reasoning used by mathematicians for that of traditional logicians, for to Galileo the greater certainty of mathematics comes not from contemplation of ideal objects (*pace* Koyré) but from use of a superior technique of reasoning. What this means in detail and how his concept differed from that of traditional Aristotelian logicians I will have to show by examining specific arguments (see, especially, Section IV). Generally speaking, however, Galileo tried to follow the deductive method of the mathematicians in developing a body of propositions rigorously derived from true and evident principles, preferably already known and widely accepted. In fact, his chief obstacle in completing the new science of motion arose precisely from the fact that it required new principles not immediately evident in the sense Galileo believed necessary. Far from using the modern method of hypothesis, deduction and experimental verification, Galileo never quite saw that the principles of mechanics, once made evident by reason and immediate experience, could still be falsified through remote consequences. Nor did he fully grasp that where immediately evident principles could not be found and less evident principles had to be used these might be adequately confirmed through systematic testing of further consequences.

But in astronomy, Galileo like many others insisted upon empirical confirmation of hypotheses together with falsification of all likely alternatives. Such a method, however, could not yield the conclusiveness of mathematics. As Galileo became more deeply involved in the struggle to establish the Copernican system as the 'true constitution of the universe,' this became a serious obstacle. If certitude in science is obtained only when true conclusions are derived from true and evident principles, how is astronomy to be made certain? It is not clear that Galileo ever believed that mathematical certainty could be achieved in astronomy, but this was in effect what the Church required before accepting the Copernican system (see Drake, 1957, p. 164) and Galileo came as close as he could to meeting this demand. After his discoveries with the telescope he increasingly insisted upon certain knowledge of observable 'facts.' But a difficulty here was that these 'facts' – usually geometrical properties observed in physical objects – were not essences, as Galileo himself recognized in calling them "particular accidents" (see Section III, below). This, of course, gave him a science dealing with 'accidents' and, in the beginning, Galileo – like the Aristotelians of his time – considered such a science to be inferior.

Thus Galileo had major methodological difficulties stemming both from his adherence to the method of the mathematicians and from the weight of the traditional Aristotelian concept of science, and although there are few explicit clues in his writings he seems to have looked for alternatives in various philosophical currents of his time, which along with Platonism also included atomism, empiricism and skepticism. At one time, for example, we find Galileo distinguishing between knowable mathematical properties and the unknowable essences of things. Elsewhere he argues from effects to causes by rules of inference which roughly correspond to Mill's methods of agreement, difference and concomitant variations, then being popularized by Francis Bacon.[3] In later writings Galileo examines the nature of mathematical reasoning, a topic of widespread interest in his time, and he attempts to interpret empirical arguments such as those employing Mill's methods in the light of such reasoning. In the end we find Galileo regarding certain non-deductive procedures involving inferences from effects to causes as 'little lower than mathematical proof' (see Section IV), and such procedures enable him to give what he considers a very nearly conclusive proof of the earth's motion. But in his mechanics, where he insisted on a strict mathematical standard of proof, he could not allow himself to rely upon the empirical methods which sufficed elsewhere, and there is far more ambiguity in the treatment of his principles in the last work on motion than is generally realized.

It was not, however, the discovery and establishment of new principles which for Galileo himself constituted his chief contribution to mechanics, and herein lies another difficulty in many analyses of his work. The main thrust of Galileo's efforts was not towards discovery of new principles but discovery of new consequences of known and accepted principles. It was the entire structure of mathematical propositions on motion and the techniques he developed for deriving them which Galileo regarded as his finest achievement. For him the most important aspect of the new science of motion was its exploratory power, or as he would put it, the capacity for 'demonstrative discovery' of new propositions. Growth of knowledge for Galileo meant primarily the generation of true propositions through rigorous demonstrations from evident principles, and this shift to greater emphasis on exploration of remote consequences, sought for their own sake, may indeed be a significant characteristic of early modern science. Although his treatise contains only a few propositions of permanent value and employs a clumsy form of mathematics which was soon superseded, it may nonetheless be an important forerunner of many later treatises in rational mechanics.

Elsewhere I have dealt in detail with Galileo's mathematical explorations (1974). Here I wish to focus primarily on his methodological procedures and assumptions and trace some of the continuities and discontinuities which run through his main writings from the earliest essay on motion, through the writings of his middle period, to the final mathematical treatise. In the end we shall find Galileo uncertain whether to present his work as a pure mathematical treatise on the one hand, or as a science of motions found in nature on the other. This ambiguity reveals the strength of his ties to ancient methods and assumptions and measures his distance from the modern world. At the same time it shows him to be a man of his own time doing work that points forward, not to the nineteenth or twentieth centuries but, as one might expect, to the eighteenth.

I. THE METHOD OF THE MATHEMATICIANS: OLD ESSAYS AND THE BEGINNINGS OF A NEW SCIENCE

Galileo's early educational background is not yet known in much detail. Some unpublished notes on logic show him to have been well schooled in Aristotelian science and method (see notes 1 and 2 above) and he is generally believed to have been introduced to mathematics only after he began his studies as a student of medicine at the University of Pisa (Settle, 1968). Subsequently, he became a lecturer on mathematics at that university, and shortly afterwards he composed an essay on motion which draws on Archimedean hydrostatics and ideas from the ancient atomists to revise Aristotle's physics of motion.[4] This project led by as yet unexplained steps into a treatment of motion along inclined planes which set Galileo on the road to his new science. The final treatise on motion was not completed for some fifty years, but even in the earliest stages we can see significant changes in his concepts of science and method as he looked away from Aristotelian science to the works of the mathematicians for better models of demonstrative science. Among other things, this meant more rigorous standards for examination of assumptions and deductive procedures and a more rationalist approach to science from which Galileo never completely escaped. The mathematical model seems also to have led to a sharper distinction, at least in practice, between strictly demonstrative sciences such as mechanics, where the purpose was to derive conclusions from evident principles, and sciences such as astronomy, where the purpose was to establish hypotheses. In this section we shall take a brief look at these initial developments.

The old essay on motion starts with a cosmological discussion in which Galileo is already departing radically from Aristotelian physics. He follows other (unnamed) philosophers who were "perhaps wrongly criticized by

Aristotle" for assuming that there is a single kind of matter in all bodies and that heavier bodies are those with "more particles of that matter in a narrower space" (Drabkin and Drake 1960, p. 15). Galileo further assumes that the total amount of matter in the universe is divided equally among the four elements, from which it becomes reasonable to suppose that earth, the most dense matter, would occupy the smallest space at the center of the world, while the other elements would occupy successively larger areas about this center according to their densities (*ibid.*). In support of this cosmology, he observes that heavy bodies move downward due to their heaviness and that light bodies will not remain at rest under the heavier ones. This property is "imposed by nature" and it is in accordance with the general arrangement (*ibid.*, p. 16). From this it follows that "natural" motion is caused by relative heaviness and lightness (Galileo intends relative density) of bodies with respect to the medium in which they are situated (*ibid.*) [4a].

Thus far, the argument is presented in purely qualitative terms. However, it is based on the hydrostatics of Archimedes and Galileo attempts to establish a rigorous foundation for his theory by presenting his own (but faulty) proofs of three theorems of Archimedes: bodies of the same density as the medium remain at rest in the medium, those of lesser density ascend and those of greater density descend (see Shea, 1972, pp. 16–24, for analysis of this material). On the basis of the hydrostatic theorems, Galileo argues that the cause of motion is relative density and goes on to develop some entirely new mathematical theorems on motion. These give the ratios of speeds for bodies of the same or different densities descending through the same or different media, and Galileo uses these results to criticize Aristotle's doctrine of motion which naively relates speed directly to weight and inversely to resistance of the medium (Drabkin and Drake, 1960, pp. 24, 26–33).

Galileo's method is explicitly modeled on mathematics and he tells us that this method "will always be to make what is said depend on what was said before, and, if possible, never to assume as true that which requires proof" (*ibid.*, p. 50). Aristotle is singled out for criticism because he took as axioms propositions "which not only are not obvious to the senses, but have never been proved, and cannot be proved because they are completely false" (*ibid.*, p. 42). We shall find this a persistent theme in Galileo's writings, one that highlights an important difference which he saw between the mathematicians and the logicians. But even the former do not entirely escape some implied criticism. For the young Renaissance scientist, the proper basis for science is to be found in such evident principles or suppositions as "that the heavier cannot be raised by the less heavy" (*ibid.*, p. 21). Thus he seeks a simpler, more evident foundation for Archimedes' hydrostatics, ignoring the latter's very difficult and hardly obvious hydrostatic principle. Later, he attempts to improve upon the Euclidean theory of proportion as well.[5]

Nowhere do we find a full discussion of Galileo's criteria for 'evident' principles, but it appears that such principles must be simple enough to be rationally evident and it must be possible to show correspondence to empirical reality through direct visual demonstration. (See Section V for Galileo's most extensive remarks on this subject and some illuminating examples.)

Now an important consequence of Galileo's explicit methodology is this: if one's foundations consist in true and evident principles, empirical confirmation of rigorously deduced consequences is unnecessary since true conclusions are necessarily derived from true premisses. This is exactly his assumption when dealing with demonstrative sciences such as hydrostatics and the mechanics of motion along inclined planes. Thus, whenever consequences derived from the premisses fail to agree with experience there must be some explanation which will save the premisses. In the old *De motu* Galileo warns his readers that if they try to find his results in experience they will fail due to various "accidental" effects (Drabkin and Drake, 1960, pp. 37–38). This is an interesting move which follows from pushing Aristotle's definition of the 'natural' to its logical conclusions. For, if the 'natural' is that which is intrinsic or inherent in things, nothing external to a moving body can affect its real 'natural' motion in an essential way. Therefore, since the motion of falling bodies is affected by the medium through which the bodies fall, the resulting motions cannot be 'natural,' and thus natural motion is no longer that which is observed in nature. It can occur only in a vacuum (*ibid.*, pp. 22n, 63, 119).

From the hydrostatic principles and theorems of *De motu* it also follows that natural motion is not accelerated. Since entirely dependent upon density it must be inherently uniform. Acceleration then is 'accidental' in some sense. Yet this accidental feature of motion is hardly accidental in the Aristotelian sense of that which is neither always nor usually so, for it occurs with considerable regularity. Nor is it accidental in a number of other senses used by Galileo to refer to "material hindrances" and deviations from an ideal form (see Koertge, 1977). In fact, acceleration appears as a kind of accident which must be accounted for through its cause rather than the kind to be dismissed as an irregularity in nature. Towards the end of his early essay on motion (in Chapter 19), Galileo gives what he considers a brilliant demonstration that acceleration in free fall is 'accidental' since it is caused by gradual decay of the impressed force which raises the body to the point from which it begins to fall. Acceleration is thus shown to be an unnatural (or violent) motion.

Here we find a 'resolutive method' being explicitly employed by Galileo as he announces that this is the method by which the cause of the effect was investigated: *huius effectus causam indagemus, haec resolutiva methodo utemur* (Drabkin and Drake, 1960, p. 88; *Opere* I, p. 318). His reasoning then proceeds through a chain of steps from the effect to be explained to a reason for this effect and hence to the reason for the reason (arguing at each step that the reason cannot be otherwise), until he arrives at an established principle – that the projecting force gradually decays – from which the entire chain will follow. From this principle it follows that at some point the projecting force will no longer be sufficient to raise the projected object, at which point the object begins to accelerate downward. Acceleration then continues until the impressed force is entirely dissipated, after which the object falls uniformly at its 'natural' speed, determined in effect by its density.

The principle found is neither discovered nor established by the resolutive method but has already been rendered evident by numerous arguments and examples (in Chapter 17). 'Discovering the true cause' thus means discovering which known principle provides the ground for the given proposition. Although evidently modeled on medieval discussions of the resolutive method, the fact that it is a known principle that is found suggests the analytic method of the Greek geometers which was employed to find *proofs* rather than the resolutive method of the medievals where the emphasis seems to be on the discovery of *principles* (Randall, 1961).

The analysis of the Greeks, already known through various sources, received fresh attention with Commandinus' translation of Pappus' *Treasury of Analysis* in 1588.[6] Galileo may already be familiar with this work which discusses the methods of analysis and synthesis (rendered as resolution and composition by Commandinus). However, the language and form used in the *De motu* are more suggestive of scholastic arguments; it is only the purpose – the search for known rather than unknown principles – which points to influence from the mathematicians. Later, however, we shall encounter the resolutive method again, but described in a manner closer to that of Pappus (but Galileo will follow Commandinus in speaking of resolution rather than analysis).

Even in his earliest use of this method, however, Galileo, although still under the influence of his scholastic education, is clearly looking to the mathematicians for his methodology. True principles are those which are clearly evident in the sense of the 'common notions' or axioms of Euclidean geometry. They are never supposed established by a posteriori argument but only through immediate observation and rational intuition. Furthermore, Galileo purports to carry out his deductions in a more rigorous fashion than that of Aristotle, not simply because he employs mathematics more often in his arguments but also because he tries to follow the mathematicians in carefully stating all of his assumptions and never assuming what he wishes to prove. Galileo was well versed in peripatetic logic, but it appears that he came fully to appreciate the nature of conclusive arguments not through the precepts of Aristotle or his medieval followers, but rather through the mathematical treatises of the ancient Greeks.

One result of Galileo's conversion by the mathematicians was an excessive rationalism, as shown above in his treatment of acceleration. (We shall see this rationalism persist, albeit in more tempered form.) Similarly, his first study of motion along inclined planes had little bearing on the behavior of physically real bodies on physically real planes and it resulted in a number of curious, unverifiable propositions (Drabkin and Drake, 1960, p. 69). Nonetheless, an important advance was made by introducing the problem of motion into an entirely new context, that of the mechanics of simple machines. This was the first step towards a radical translation of the science of motion from an Aristotelian physics to mechanics in the Archimedean-Jordanian tradition (see Wisan, 1974, section 2). The next step appears to be a deeper examination of the mechanics of simple machines, and shortly afterwards we find Galileo working seriously on a science dealing entirely with the 'accidents' of accelerated motion, where these accidents are those which regularly occur in nature (*ibid*.).

Still, as early as 1602, we find Galileo writing to his patron Guidobaldo del Monte that there can be no certain mathematical science dealing with material things (*Opere* X, p. 100) and, as we shall see in Section V of this paper, it is by no means evident that Galileo ever came to a clear and firm conclusion about the relation between his propositions on motion and the physical situation they described. Although he claimed to have confirmed the times-squared law ($t^2 : s$) by more or less direct experiment, the more fundamental principles of the treatise on motion seem to have lacked such direct confirmation. So did the more remote results. Indeed, it does not seem to have occurred to Galileo that remote consequences could have any bearing on the validity of principles of a demonstrative science (see Sections II and V, below). From the beginning of his work on motion, Galileo sought "totally indubitable principles" from which to derive his propositions. These are his words in a letter of 1604 to Paolo Sarpi where he introduces as very natural and evident (*ha molto del naturale et dell'evidente*) his mistaken law ($v : s$) that in free fall velocity acquired is proportional to distance fallen (*Opere* X, p. 115). But he also indicates that he thinks he can demonstrate this proposition and to the very end Galileo vainly sought direct confirmation of those propositions on which he explicitly based the treatise on motion, that is, the correct law of fall ($v : t$), that acquired velocity is proportional to time of fall, and the postulate that bodies acquire equal speeds in descending through equal vertical distances (see Section V, below for fuller discussion; see also note 35).

Recent studies of previously untranscribed notes from Galileo's manuscript relating to the work on motion reveal definite evidence of real experiments and suggest that such experiments played a more important role in the development of the new science than recently thought by most historians. Certainly it appears that Koyré's extreme position on this matter – that experimental results played virtually no part in the development of Galileo's thought – is quite untenable.[7] We must take care, however, that we do not now go too far in the opposite direction. These notes which contain little besides diagrams, calculations, and suggestive sets of figures, must be carefully studied along with other manuscript notes that reveal more clearly Galileo's theoretical concerns. One set of notes, for example, which probably relate to the realization that acceleration must be regarded as 'natural,' exhibit some puzzles about the relation between uniform and accelerated motion which Galileo never fully resolved (Wisan, 1974, pp. 200–204). Another set includes demonstrations of the times-squared theorem and the double-distance rule, carried out in a crude manner reminiscent of methods in fourteenth century kinematics (*ibid.*, pp. 204–210) but based on the mistaken law of fall ($v : s$).

Then, there is a fragment in which Galileo appears to be examining the consequences of assuming acquired velocity proportional to time of fall ($v:s$) and also to the square root of the distance fallen ($v:s^{1/2}$). However, his demonstration from these assumptions seems to lead to inconsistent results (*ibid.*, pp. 210–215), and subsequent fragments which can be linked with a letter of 1609 show Galileo moving away from his first attempt at a kinematic foundation towards a modified statics based upon what appear to be purely rational principles (*ibid.*, pp. 215–229).

The recently discovered evidence of experimentation must be interpreted in light of this theoretical background. For example, in the case of folio 116v, we find evidence of an experiment carried out by dropping an object from different vertical heights (or possibly rolling it down inclined planes of different heights) and then deflected along a horizontal. What appear to be experimental results are marked at five points along a horizontal line in a diagram which shows trajectories to these points from what is indicated to be a table top. Then, *after* the experimental results were recorded, calculations were made which assume (1) $v:s^{1/2}$ and (2) conservation of motion along the horizontal (no one investigating this fragment seems to have noticed the obvious temporal order of the results as Galileo wrote them down). Those who have closely examined the data and reproduced the experiment claim a very good fit between calculated and experimental results, but they disagree about what Galileo intended to test: the principle of horizontal inertia (Drake, 1973, 1975a), the times-squared law (Naylor, 1974), or the proportionality $v:s$ (Costabel, 1975).

Space does not permit detailed discussion here of the data and calculations, or of the difficulties in the various interpretations offered. However, a plausible interpretation which takes into account theoretical as well as experimental notes, might be as follows. Galileo set out to test the mistaken law of fall, $v:s$. This obviously failed to satisfy Galileo (contrary to Costabel). If he then turned to the simplest alternative, $v:s^{1/2}$, this would account for the recording of experimental results before making calculations, since the simpler hypothesis required none. Furthermore, if we suppose this experiment made after Galileo's letter to Sarpi (*Opere* X, pp. 115–116), in which he claims to derive the times-squared law from the erroneous law of fall ($v:s$), the performance of such a test would be consistent with Galileo's remark on folio 128 (where the argument is given in detail) that he thought he might be able to demonstrate that principle (*Opere* VIII, p. 373). The test, of course, failed, showing $v:s^{1/2}$ instead. Meanwhile, however, as shown by Drake (1975b), Galileo may have had an independent experimental demonstration of his law of odd numbers, from which the times-squared law derives immediately. Thus, when the experiment on folio 116 failed to show $v:s$, but suggested $v^2:s$ instead, the latter combined with $t^2:s$ would give the correct law of fall, $v:t$.

At first, this may well have led to confusion. Galileo's argument for $t^2:s$ (and his double-distance rule as well) depended on $v:s$, thus creating another puzzle about accelerated motion (Wisan, 1974, pp. 199–210). Moreover, the postulate that equal speeds are acquired in descent through equal vertical distances seems to have been identified, at least for a while, with the principle $v:s$ (*ibid.*, p. 219). It can be shown that these fundamental problems were not quickly solved (*ibid.*, pp. 220–229). My guess is that Galileo took the results of his experiments giving $v^2:s$ and $t^2:s$, derived $v:t$, and began an investigation on folio 152r to check the consistency of these results with the double-distance rule by using a method which drew on medieval kinematic techniques. When this failed due to a conceptual error, Galileo then abandoned the kinematic approach and returned to his theoretical foundation in the mechanics of the inclined plane, as shown in fragments related to correspondence with Valerio in 1609 (*ibid.*). The erroneous law of fall, $v:s$, continued to be employed and Galileo clearly retained confidence in both the times-squared law and the postulate of equal speeds. The principle $v:t$ was needed for the book on projectile motion, which Galileo seems to have put aside for a while, but the times-squared theorem and the postulate were sufficient foundation for most of the book on accelerated motion and both were probably well enough confirmed by experience for him to proceed with his research.[8] But this was not sufficient for the final presentation of a new science. The times-squared theorem was not yet proved mathematically to Galileo's satisfaction (Wisan, 1974, p. 221) and it was not a rationally evident principle.

Thus, Galileo remained uncertain and dissatisfied with the foundations of his new science for a considerable period. Indeed, he may not have gotten his final derivations of the times-squared law and the double-distance rule until after a return to medieval sources which showed how to do this with $v:t$ (*ibid.*, section 8). As I hope to show in greater detail, Galileo sought neither purely rational principles nor principles that could be verified indirectly. What he wanted was rational principles which were immediately evident through both reason and experience. As we shall see, however, this was an impossible ideal. Galileo had eventually to resort to other means, and his ambiguous language served at once to point in two quite different directions. But more about this later, for, in any event, after having worked out a few propositions on projectile motion and started some new investigations of accelerated motion, he turned his attention to astronomy.

Throughout Galileo's study of motion, as we shall see, he resists depending on indirect confirmation of hypotheses. But during the same period in which he was beginning this study, he composed an essay on cosmography which shows an altogether different concept of method (*Opere* II, pp. 211–255). His method in astronomy was not that of the mathematician who derives true conclusions from true principles but that of deriving conclusions from hypotheses, confirming the conclusions by observations and showing that all alternative hypotheses which are proposed must fail. In the *Cosmographia*, Galileo assumes Ptolemaic astronomy and Aristotelian cosmology, but by 1610, after his telescopic observations, he openly espoused the Copernican system and began a vigorous campaign for acceptance of this system as the 'true constitution' of the universe. No hypothesis, of course, can be established with absolute certainty unless all possible alternatives can be eliminated, and this was the main technicality on which the Church refused to accept the Copernican system. Galileo was aware of the force of this argument, and in his earlier *Cosmographia* he did not suggest that his demonstrations were necessary in a mathematical sense, even though he seemed to take for granted that it was the real structure of the universe about which he was writing. But as he became convinced of the truth of the Copernican system his methodological approach changed. For the telescopic observations decisively swept away the Aristotelian cosmology and seemed to provide immediate evidence that Venus revolved about the sun (see Section III, below). Perhaps for this reason Galileo tended to present such observations as though they were direct proofs of the Copernican system rather than mere confirmations of the hypothesis. If so, it should not be surprising to find him paying more attention to empirical theories of science and method, and this appears to be the case.

Unfortunately, it is difficult to determine exactly what Galileo's real views on method were, particularly in the period after 1616 when teaching the

Copernican system was prohibited. Especially in the *Dialogue* Galileo's procedures are often obscured by polemical purposes and the need to dissimulate. However, there are a number of important clues in some of his writings before 1616, and it will be useful to look at the *Discourse on Bodies in Water* and then the *Letters on Sunspots*, both of which were written during the same few years between 1610 and 1613.

II. THE MATHEMATICAL MODEL AND NEW EMPIRICAL METHODS: DISCOURSE ON BODIES IN WATER

Galileo's treatise on floating bodies, published in 1612 (*Opere* IV, pp. 63–140; translated in Salusbury, 1967), is rich in linguistic and procedural clues to a new approach to science. Here for the first time we find him employing new methods which will later be used as rules of inference for deriving causes from effects and it will be through such rules that empirical arguments eventually acquire for him a force nearly equal to that of mathematics. On the other hand, hydrostatics remains for Galileo a strictly demonstrative science in the sense of the old *De motu*. He makes more use of experiments in supporting his theory, and a fundamental principle which is less than immediately evident is established by extensive argument which appeals to a number of experiments. But once the principles are established, failure to confirm remote consequences has no effect. As we shall see here and elsewhere, this ambiguity arises from an inevitable conflict between the ideal of a demonstrative science modeled on Euclidean geometry and the difficulty in satisfying this ideal as soon as one attempts to go beyond the most elementary kind of mechanical theory. In this section, then, we shall see the emergence of new methods and attitudes which will nonetheless remain – even to the very end – within the old framework.

The *Discourse on Bodies in Water* is a rambling discussion which grew out of controversy over the question, why do bodies float or sink in water? (See Shea, 1972 for the history of this argument and for detailed analysis of some of Galileo's propositions.) Galileo reviews the dispute and states his 'principal proposition' which is that bodies descend or ascend in water according to whether they are denser or less dense than water (Salusbury, 1960, pp. 4–5, 47). This, he asserts, is the "true, intrinsecall, and totall Cause, of the ascending of some Sollid Bodyes in the Water, and therein floating; or on the contrary, of their sinking" (*ibid.*, p. 4). To disprove this proposition, his opponents had produced a thin board of ebony which was heavier than water but which floated and they claimed that this experiment proved their own theory that it is the shape of the body which determines whether or not it will sink or "swim" (*ibid.*, p. 26).

The experiment with the board of ebony was an embarrassment to Galileo and his treatise was written in order to demolish the force of the experiment by means of counter experiments and a general mathematical theory which provided an explanation for the apparent anomaly of the ebony board that floats. Galileo's treatise then is in part a deductive treatment of hydrostatics and in part an extensive discussion of various experiments which are offered both as counterexamples to the conclusions of his adversaries and as positive support for his own view.

As in the old *De motu*, Galileo begins his demonstrations with evident principles and attempts to derive rigorous conclusions by using mathematics. He tells us that since "demonstrative Order *[progressione dimostrativa]* so requires, I shall define certain Termes and afterwards explain some Propositions, of which, as of things true and obvious, I may make use of to my present purpose" (*ibid.*, p. 5). And again, as in the old essay, he does not employ the hydrostatic principle of Archimedes but

I, with a different Method, and by other meanes, will endeavour to demonstrate the same [theorems], reducing the Causes of such Effects to more intrinsecall and immediate Principles, in which also are discovered the Causes of some admirable and almost incredible Accidents . . . (*Ibid.*)

Galileo then defines certain critical terms, takes some principles from "Mechanicks," and derives several theorems on hydrostatics. The first few theorems include the basic hydrostatic theorems of Archimedes from which Galileo's 'principal proposition' derives as a general statement of the results. Although this result is rigorously deduced from a foundation in mechanics, and one which is actually more Aristotelian than Archimedean, as Galileo himself makes clear, he is nonetheless obliged by his critics to further defend his proposition by showing why the experiment with the board of ebony does not falsify it. This leads into a long discussion, labeled 'Corollary III' to Theorem IV in Salusbury's translation, of some criticisms of Archimedes. Then, under 'Theorem V' (Salusbury's label is again misleading; however, I use his labels for convenient reference) Galileo gives a detailed discussion of experiments which will disprove the contention of his opponents that different shapes account for the sinking or floating of objects and provide the foundation for his own explanation of the experiment with the piece of ebony. Here Galileo, for the first time in his writings, employs the methods of agreement and difference, which he introduces by remarking that he will proceed by "a more excellent Method" (*modo più esquisito*) than that of his opponents (Salusbury, 1960, p.29; *Opere* IV, p. 91). He then describes an experiment in which he takes a ball of wax which is lighter than water and he molds it into various shapes, none of which sink. But if a bit of lead is added, the wax will sink, whatever its shape, and it rises again when the lead is removed.

Galileo then attempts to explain the phenomenon of the floating board of ebony. He points out that if the board is place "into" the water rather than "upon" the water, it will sink (Salusbury, 1960, p. 32). That is, if the top of the board remains dry it will float, but if the board is completely wet, it will sink. Thus it must be something other than the shape that causes the board to float, and Galileo finds this cause in a principle of affinity between solid substances and the air. When the board of ebony is placed on the water it forms, in effect, an object composed of both ebony and air. For, as Galileo observes, the board sinks slightly below the surface of the water. Thus, the board, together with air contained within the 'ramparts' of water formed around the board, constitutes an object which is lighter than water and which therefore floats just as an empty brass kettle will float. The board will sink only when its top is made wet, for this breaks the "contiguity" between the ebony and the air (*ibid.*, pp. 36–39).

Galileo argues at length for his principle of affinity, citing many observations and experiments, both as direct confirmations of the affinity and as counterexamples to possible objections. In other words, Galileo argues for a new hydrostatic principle, appealing to experiment for more or less immediate confirmation, or, as Galileo puts it, to make his principle 'manifest.' For him, failure to confirm remote consequences will not affect the truth of his principle of affinity nor cast doubt upon his 'principal proposition.' Indeed, it would appear that the mere fertility of the theory proved its truth. At the end of 'Theorem V' Galileo tells us that

> since we have found the true Cause of the Natation of those Bodies, which otherwise, as being graver than the Water, would descend to the bottom, I think, that for the perfect and distinct knowledge [*intera e distinta cognizion*] of this business, it would be good to

proceed in a way of discovering demonstratively [*d'andar dimostrativamente scoprendo*] those particular Accidents that do attend these effects . . . (*Ibid.*, p. 45)

There follow a number of mathematical demonstrations which develop the consequences of his theory and show how it works for objects of different shapes and densities.

Thus, Galileo proceeds to employ deductive demonstration as a means of discovering new 'accidents.' Here, he seems to be drawing a deliberate contrast between his method, which he claims to be one of demonstrative discovery, and the resolutive method of Zabarella. Discovery is not to be made by reasoning from experience to unknown principles, but rather by deducing accidents from their 'true causes.' On the other hand, Galileo's remark that "a perfect and distinct knowledge of this business" is to be aided in some way by the discovery of the accidents may contain an echo of Zabarella's doctrine.[9] But Galileo's intention is to derive as a theorem a proposition which will account for the apparently anomalous experiment with the board of ebony and, as we shall see, he actually derives 'accidents' which cannot be confirmed in experience. He does not however see this as an indication that he may have less than perfect and distinct knowledge of the matter. He simply offers yet another experiment which seems more or less directly to confirm his principle.

The difficulties with his theory emerge as a result of Theorem VI and its corollaries (Theorems VII and VIII) which establish the conditions under which a body denser than water will float. For the thickness of the object (but not its length or breadth) turns out to be critical. But according to the theory neither size nor shape can be relevant and he sidesteps this difficulty through a vague concept of the 'possible altitude' of the ramparts formed by the water.[10] This factor, introduced without comment, and never treated mathematically, is regarded as having no bearing on the true cause of bodies sinking or swimming. It simply explains why most heavy objects, however carefully they are placed in water, do get their tops wet. Galileo is now ready to account for the experiment with the board of ebony. The answer to the experiment comes from Theorem XIII: all objects (heavier than water) sink or swim depending on whether or not their tops become wet when they are placed in water. And this, he affirms, simply reduces to his proposition that "the true, Naturall and primary cause of Natation or Submersion" is the difference in density of the "solid Magnitude" (the object together with the contiguous air, or the object alone if contiguity is broken) and that of the water (*ibid.*).

First Galileo argued against his opponents on the basis of experiments and counterexamples. Now the 'effect' which was used against him has been derived from its 'true cause' – Galileo's principal proposition. But his theory is vacuous. It cannot be tested, and Galileo knows this. In fact, he tells us that "if one should take a Plate of Lead... a finger thick, and an handfull broad... and should attempt to make it swimme, with putting it lightly on the water, he would lose his Labour" (*ibid.*). For "if it should be depressed one Hairs breadth beyond the possible altitude of the Ramparts of water, it would dive and sink" (*ibid.*, pp. 58–59). The theory can only be verified by taking a piece of lead so thin that "a very small height of Rampart would suffice to contain so much Air, as might keep it afloat" (*ibid.*, p. 59). That it is in principle impossible to test his mathematical theory in other cases does not suggest to Galileo that the theory might be false and, consequently, that his principle might need modification. Instead of reexamining his principles or investigating the meaning of 'possible altitudes,' Galileo simply offers a final experiment which seems directly to confirm the supposed 'affinity,' and he cheerfully concludes that "I have fully demonstrated the truth of my Proposition"[11] It would appear, then, that the truth of a mathematical theory does not depend upon verification of its remote consequences.

In Galileo's science of motion, we shall see a similar attitude towards remote consequences of mathematical principles. But in astronomy the situation is altogether different. The consequences of hypotheses must be fully confirmed, however remote. But some of these consequences will appear more and more as 'facts' which seem to be known in the same direct way in which the principles of mechanics are known. Thus, although the distinction between astronomy and a demonstrative science remains, the distinction between empirical facts and rational principles diminishes. As I have already suggested, the rules for inferring causes from effects will also play a part in the merging of empirical and rational arguments, and in a later essay – the *Saggiatore* – Galileo speaks of such rules as "laws of logic."[12] Finally, in the *Dialogue* we will see these 'laws of logic' used to arrive at important results. First, however, we should examine the *Letters on Sunspots*, written about the same time as the *Discourse on Bodies in Water*, where Galileo stresses accuracy of observation and explicity distinguishes between what we can know through direct experience and that which we know by reasoning.

III. FROM RATIONALISM TO EMPIRICISM TO 'MITIGATED' SKEPTICISM: LETTERS ON SUNSPORTS

During the same period in which Galileo was writing the discourse on the behavior of bodies in water and ignoring experimental results which could raise questions concerning his theory, he was also improving his methods of telescopic observation and carefully interpreting results in a manner which conformed as strictly as possible to observational data.[13] While the *Discourse* was in press, he began writing letters concerning his observations of sunspots.[14] In these letters he continually criticized those who were attempting to explain away observational data so as to salvage their a priori theories concerning the nature of the sun (Drake, 1957, pp. 118, 140–142). Galileo shows not the slightest awareness that he too has salvaged an a prior theory in the same way. Nor should one expect this. For his mechanics and hydrostatics were deduced from 'true and indubitable' principles, whereas in astronomy there were no immediately evident principles. However, he was finding that in astronomy there can be very accurate (*accuratissime*) observations and that conclusions which are more than merely probable can be inferred from them by use of mathematics. This is the main methodological message of the letters concerning sunspots.

The letters were written to Mark Welser of Augsburg, who had sent Galileo, for his comments, an anonymous treatise written by the Jesuit Christopher Scheiner (Drake, 1957, p. 81). Like other peripatetics of the time, Scheiner was willing to make observations with the new telescope and to revise many of his opinions. But that the sun itself was the seat of a process of generation and corruption was more than most of them could accept. From the general principle that the sun must be inalterable, they deduced that the sunspots could not be on the body of the sun. This, of course, was the sort of a prioristic reasoning to which Galileo most vigorously objected. Since we cannot directly examine the sun, its inalterability cannot be taken as an evident principle. Similarly, since sunspots are not accessible to close inspection, knowledge about them can only be gained indirectly as in astronomy generally. That is, one must make hypotheses, deduce consequences from these hypotheses and confirm the results through experience and observation.

Scheiner, in his treatise, argued for the hypothesis that the sunspots were tiny stars revolving around the sun. In Galileo's first letter to Welser, he set forth no specific view of his own, but merely exposed difficulties in Scheiner's arguments. In a second letter Galileo offered a proof that sunspots are either on the surface of the sun or "separated by an interval so small as to be quite imperceptible" (*ibid*., p. 106). His method of demonstration is that which is usual in astronomy. After a number of elegant demonstrations showing how various observed phenomena, such as changes of shape and degrees of separation between sunspots, agree with his own hypothesis, he briefly considers and refutes several alternative theories. Then, after citing further observations from different places on the earth which provided even richer confirmation of his own theory, Galileo rejects what he believes to be the only possible alternative hypothesis which could fit this data, that of a "little sphere somewhere between us and the sun" which could carry the sunspots (*ibid*., p. 111). But this hypothesis fails because in such a case we should see spots moving in both directions under the sun (due to the supposed separation between the spots and the sun's surface) which does not happen. The last plausible alternative is thus removed. Galileo completes his argument with the comment that "it would be a waste of time to attack every other conceivable theory" (*ibid*.). So far as he is concerned this is a sufficient demonstration for his hypothesis of the contiguity of sunspots. Even though he has not, in fact, refuted every *possible* hypothesis, he has responded to every one which he believes could reasonably be proposed, and he concludes that his hypothesis is true because

> ...all the phenomena in these observations agree exactly with the spots' being contiguous to the surface of the sun, and with this surface being spherical rather than any other shape, and with their being carried around by the rotation of the sun itself, so the same phenomena are opposed to every other theory that may be proposed to explain them. (*Ibid*., p. 109)

Here is an important instance in which Galileo has developed an elegant mathematical model to account for a physical phenomenon. No doubt he found the nice fit of his model to be quite persuasive; however, this constitutes no part of the proof. The mere fact that the phenomena can be mathematically interpreted is no guarantee of the truth of the interpretation.[15] In fact, Galileo is careful here not to claim that he has proved that the spots are actually on the sun, as this cannot be *known* without more direct empirical evidence. But we do find the claim that direct and evident experience, when available, does give certain knowledge. At the end of his second letter Galileo associates himself with the Aristotelian view that the best philosophy is that which depends upon the evidence of the senses and points out that he contradicts

> the doctrine of Aristotle much less than do those people who still want to keep the sky inalterable; for I am sure that he never took its inalterability to be as certain as the fact that all human reasoning must be placed second to *evident experience*. Hence they will philosophize better who give assent to propositions that depend upon manifest observations, than they who persist in opinions repugnant to the senses and supported only by probable reasons. (*Ibid.*, p. 118; my emphasis.)

'Evident experience' and 'manifest observations' thus give rise to conclusions which are not merely probable but are certain in some sense. There is additional evidence for this view in the same letter where Galileo distinguishes between those things which are 'manifest' to sense and do not require use of reason (*discorso*) and those things that reason deduces and concludes from "certain particular accidents provided by *sensate osservazioni*" (*ibid.*, p. 107; *Opere* V, p. 117). In the first category are such properties as different densities and degrees of darkness, changes of shape, and movements. Galileo argues, for example, that the revolution of Venus about the sun is indubitably demonstrated by a "single experience. . . namely, our seeing Venus vary in shape as does the moon" (Drake, 1957, p. 133). For Galileo apparently, there is nothing hypothetical about the revolution of Venus, but it follows with "absolute necessity" from observation with the telescope (*ibid.*, p. 94; see also Drake, 1967, p. 322 and *Opere* VII, p. 350). But that the sunspots are actually in contact with or imperceptibly distant from the sun and are moved by its rotation is not directly knowable and can be inferred only by reasoning mathematically from observable properties and verifying the conclusions.

Galileo's distinction is really between low level generalizations of a kind which often pass as 'facts' and those which more clearly involve inferences. Here again this sounds very much like a naive empiricism. 'Necessary' truths are learned through direct observation, and we sometimes find him 'deducing' (*deduca*) from observations (Drake, 1957, p. 107; *Opere* V, p. 117). Galileo does not elaborate on how he is using these terms, but throughout the letters on sunspots and in the later *Dialogue* he frequently contrasts the knowable fact with the less knowable generalization and insists on the primacy of the former. For Galileo, knowledge in astronomy consists mostly in *facts* known with certainty through observation and experience. And knowledge of such facts becomes more and more comparable to knowledge of fundamental principles as in mechanics. That is, both seem to be certain knowledge of the physical world acquired in some way through direct experience.

The arguments and language in the letters on sunspots suggest a radically different philosophical orientation from that which underlies the earlier *Cosmografia*. The Aristotelian cosmology, of course, is no longer assumed, nor is Ptolemaic astronomy. Also, whereas in the *Cosmographia* there is reference to knowledge of substance (*Opere* II, p. 212), such knowledge is now denied. At the same time, however, there is a new insistence on 'absolutely necessary' conclusions from empirical arguments. Thus, there appears to be a growing element of empiricism. Yet, along with these new claims for empirical knowledge and in spite of insistence on our ability to know the 'true constitution of the universe,' Galileo seems to have turned at least a few degrees towards philosophical skepticism. Such a position does not necessarily conflict either with his empiricism or with his rationalism, but both must be understood within limits defined in the second letter on sunspots. In this letter he adopts a frankly skeptical view of our ability to know the "true and intrinsic essence" of natural substances, being satisfied to know only some of their properties, or *affezioni* (Drake, 1957, p. 124). These properties turn out to be, for the most part, those mathematical properties which Galileo will later refer to as "real accidents." [16] He says that "although it may be vain to seek to determine the true substance of the sunspots, still it does not follow that we cannot know some properties of them, such as their location, motion, shape, size, opacity, mutability, generation, and dissolution" (Drake, 1957, p. 124). These properties may still be "the means by which we shall be able to philosophize better about *other and more* controversial qualities of natural substances" (*ibid.*, my emphasis). But no longer, as in the earlier essay on astronomy, does the knowledge of accidents lead to *knowledge* of substances (*Opere* II, p. 212), but only to a tentative and limited insight.

These changes in attitude towards what we can know suggest that Galileo is fishing in various philosophical currents of the time as he discards his initial Aristotelian orientation on theory of science and method. Certainly, with Aristotelian essences an inconvenience in both astronomy and mechanics Galileo could hardly have failed to notice arguments such as those flowing from the Pyrrhonean controversy. Even the 'true constitution of the universe' for which he is now undertaking to do battle may never be more than a unique arrangement of parts, involving no claims about the essential nature of "true being"[17] Indeed, the flexible Tuscan may at this time be moving towards a 'mitigated skepticism,' somewhat like that later developed by Gassendi and Mersenne, and which leaves ample room for a positive (or *resoluto*, to use his term) attitude towards both mathematical and empirical knowledge while avoiding the problem of essences.[18]

Meanwhile Galileo became more deeply involved in controversy over the Copernican system, especially as it touched upon religious issues. He was increasingly under attack for going against the Bible in holding that the earth moves rather than the sun, and in 1616 he was forbidden to teach or write anything more in defense of the Copernican system. The importance of this prohibition on Galileo's scientific writings cannot be too strongly emphasized. For it is difficult to be sure of his methodological and philosophical views from this time on. His next move was to publicly adopt a radical skepticism with regard to our knowledge of the construction of the universe. That is, he professed to doubt not only knowledge of essences but also of mathematical structures, or precisely those properties of things, knowledge of which was claimed in the second letter on sunspots. Perhaps the earliest hint of this new position is found in Mario Guiducci's *Discourse on the Comets*, published in 1619 (translated by Drake in Drake and O'Malley, 1960). Guiducci, who was a student of Galileo's and whose *Discourse* was strongly influenced if not actually composed by the master himself, asserts that "we still lack firm knowledge of the universe and must be content with conjectures here among the shadows until we know the true constitution (Drake and O'Malley, 1960, p. 57). In this passage there is an explicit reference to Seneca and the need to suspend judgment, but the reference to Plato's allegory of the cave is unmistakeable. This remark foreshadows extensive use of Platonic devices throughout most of the later *Dialogue*. We shall next examine that work and try to find our way through the wily Tuscan's devious logical games with his Aristotelian opponents and his use of Socratic irony and professions of ignorance by which he will lightly veil his real beliefs.

IV. THE MERGING OF MATHEMATICAL AND EMPIRICAL REASONING: THE DIALOGUE

Galileo believed that it was his telescopic observations which decisively established the Copernican system, and his devotion to that cause may have been motivated in no small part by personal ambition.[19] For Galileo, as for most scientists, proving an hypothesis is an achievement of higher order than putting forth a speculation, however plausibly argued. Thus, while extravagantly praising Copernicus for making his reason 'conquer sense' and marvelling over the way in which he "resolutely continued to affirm what sensible experience seemed to contradict" (Drake, 1967, p. 339), Galileo nonetheless makes it clear that it is he who has established the truth of the new system by arguments based on observations with his telescope.[20] But these observations could all be satisfied by Tycho Brahe's system in which the planets revolve about the sun while the sun revolves about an immobile earth. Therefore, more important perhaps than arguments from telescopic observations was Galileo's attempted demonstration of the earth's motion from the tides. This demonstration, he seems to have felt, came very close to establishing the Copernican system with the mathematical certainty required by the Church. However, due to the new restrictions imposed by the Church, the argument could not be made in a straightforward manner.

It was by promising to give an impartial presentation of the two main astronomical systems – that is, the Ptolemaic and Copernican – that Galileo obtained permission to publish his great *Dialogue*. The three participants are Salviati, who gives the arguments for Copernicus and who generally speaks for Galileo (referred to throughout as an unnamed 'Academician'); Simplicio, staunch defender of the Aristotelian-Ptolemaic system of natural philosophy and astronomy; and Sagredo, who presumably represents the impartial spectator. In the first day of the *Dialogue*, Simplicio assumes, on the basis of Aristotle's physics, that there are necessary reasons why the earth cannot move. Salviati responds by attacking the physical principles and logical arguments on which the Aristotelians based their system of the world. Then, later, he draws on Galileo's telescopic observations of the moon to further undermine the Aristotelian cosmos. In the second and third days, Salviati shows that the earth's daily and annual motions are consistent with naked eye observations and further supported if not absolutely proved by numerous observations with Galileo's telescope. By the end of the third day Simplicio is in retreat to the weaker position that while there may not be necessary reasons why the earth cannot move, it cannot be argued that the earth does necessarily move.[21] In the fourth day, Salviati presents Galileo's argument that tides could not occur naturally on a motionless earth, and that it is the combination of the earth's various motions that causes the tides. Simplicio then withdraws to the last line of defense: even if the tides could not be produced naturally if the earth were motionless, God could cause them by a miracle. Salviati assents to this argument, but since throughout the dialogue, it has been agreed that God, like Aristotle's nature, acts in the simplest way and would never perform an unnecessary miracle, the Aristotelian has lost the debate. (The Pope, of course, recognizing that it was his own argument which was put into the mouth of Simplicio, exercised his own arbitrary powers, and Galileo lost his freedom.)

The *Dialogue* is among other things a literary work. Galileo makes skillful but sometimes specious use of Aristotelian principles and logic, setting many a snare for unwary Simplicios, and he also uses various Platonic devices, such as Socratic irony and professions of ignorance, to veil his own thought. In the first three days of the dialogue he employs Socratic questioning and sometimes, apparently, the Platonic doctrine of reminiscence.[22] But in the last day of the dialogue there is a noticeable change in style. References to Plato and use of Socratic questioning disappear and Salviati, who has promised to do so earlier (Drake, 1967, p. 131), seems to drop his mask here, for he argues directly and urgently for the proof from the tides. Then, at the very end, he resumes his mask and the discussion ends on such an obviously false note as to deceive scarcely anyone.

The message of the *Dialogue* is clear. The traditional cosmology – an elaborate system in which Ptolemaic astronomy and Aristotelian physics blended into a comprehensive and unified whole – is false. It must be replaced by a new cosmology in which it is the Copernican astronomy which fits, albeit in less comprehensive manner, a new but incomplete physics. But in his presentation of the new system, Galileo is not only hampered by restrictions imposed by the Church but even more by methodological difficulties. Without true and evident principles from which to derive the motions of the earth and planets, there can be no certainty except that which comes from observation.

Galileo's strategy is, first, to undermine Aristotle's physics by demonstrating that his principles are not evident and that his proofs are not rigorous. Then there is a shift in emphasis as Galileo begins to focus on evidence from observations. For in astronomy it is not by deduction from general principles that one is to be persuaded, but through appeal to "specific arguments, experiments, and observations" (*ibid.*, p. 47). As in the *Letters on Sunspots*, Galileo emphasizes the primacy of the particular fact over the abstract generalization, and this theme is sounded throughout the *Dialogue*. In the second day, where Salviati presents his case for the motion of the earth, he begins with arguments from the principles of simplicity, economy, and order. But these principles, he tells us, are not set forth

as inviolable laws, but merely as plausible reasons. For I understand very well that one single experiment or conclusive proof to the contrary would suffice to overthrow both these and a great many other probable arguments. (*Ibid.*, p. 122.)

Metaphysical principles, such as the principle of order, can never be more than guiding precepts, subject to modification on the basis of experience. Only principles such as those of mathematics and mechanics which can be immediately established through reason and experience are to be taken as true, evident principles, and only from such principles does reason lead to certainty. But in the case of astronomy Galileo must convince his readers by reasoning from observation and experience, and we shall see him more and more associating such reasoning with that of the mathematicians.

In the first day of the *Dialogue* Salviati limits certain and sure knowledge to that of pure mathematics, and he even claims that "the truth of the knowledge which is given by mathematical proofs. . . is the same that Divine wisdom recognizes" (*ibid.*, p. 103). That is, there are propositions which the human intellect can understand as perfectly as the Divine intellect itself. Although God knows infinitely more propositions, those few which are understood by man are grasped with the same certainty, for the human intellect "succeeds in understanding necessity, beyond which there can be no greater sureness" (*ibid.*). But whereas the Divine mind operates through a simple intuition, our reasoning must proceed step by step from one conclusion to another. It is essentially this necessity for proceeding step by step that differentiates human from Divine understanding of a mathematical proof. Yet even here the human mind can approach the Divine in its results when it can master

> some conclusions and get them so firmly established and so readily in our possession that we can run over them very rapidly. (*Ibid*., p. 104.)

Salviati's example of this is the Pythagorean theorem, one of the more complex propositions in Book I of Euclid's *Elements*. This theorem (Proposition 47) is proved from axioms and postulates through a number of theorems related to the central developments in Book I, that is, the construction and properties of triangles, the congruence theorems, the theory of parallels and treatment of equality in the case of non-congruent areas. Galileo's concept of 'mastery' of Proposition 47 of Euclid entails that all of this be fully understood, as appears from Salviati's remark:

> For, after all, what more is there to the square on the hypotenuse being equal to the squares on the other two sides, than the equality of two parallelograms on a common base and between parallel lines? And is this not ultimately the same as the equality of two surfaces which when superimposed are not increased, but are enlcosed within the same boundaries?[23]

In other words, man's understanding of a mathematical proposition is similar to that of God's when he has understood not only the formal proof but also the underlying rationale, in this case the way in which, given the theory of parallels, the equality of non-congruent areas can be shown through reduction to congruence. Finally, the whole must be grasped in the same "continuous and uninterrupted process of thought" prescribed by Descartes' Rule VII for the "consummation of knowledge."[24]

For Galileo, of course, the rationale for geometrical proofs cannot lie in a formalized logic. No one had had much success in reducing mathematics to syllogistic logic and little progress had yet been made towards modern mathematical logic.[25] Thus, like Descartes and others he seems to have found the secret of rigorous reasoning in mathematics in the process by which one proceeds from one small step to another where each step follows so clearly and distinctly and with such apparent necessity that it cannot be otherwise. But this suggests that with sufficient care one might obtain similar success in reasoning from empirical data, and indeed Galileo (like Descartes) seems to have drawn this conclusion, albeit not without qualification. We shall examine a clear instance of this in the *Dialogue*, but first we must turn from Galileo's remarks on synthetic proof, as in Euclid, to analysis as defined by Pappus.

We have already examined one explicit discussion of a resolutive method. This was in the old *De motu* and although Galileo used the method to find a proof, as in mathematics, rather than to discover principles, the logical structure is much the same as that found in scholastic writings. Assuming an effect which was to be explained, Galileo looked for an immediate cause from which it could be derived, then for the cause of the immediate cause, and so on until the true cause was discovered, after which the effect to be explained could be rigorously derived from its true cause. This form of the resolutive method, common in natural philosophy, easily translates into a mathematical method of analysis (and indeed the former term was at first used to translate the latter) if one speaks not of causes to be discovered, but of theorems to be proved or problems to be solved.

In *Euclid* XIII there is a definition of analysis (generally regarded as an interpolation) in which one begins with the theorem to be proved and derives consequences until some known truth is encountered (Heath, 1926, III, p. 442). Then in Pappus' *Treasury of Analysis* a definition is given which seems to contain elements from traditions that stem from both mathematics and natural philosophy:

> Now analysis is the way from what is sought – as if it were admitted – through its concomitants /the usual translation reads: consequences/ in order to something admitted in synthesis. For in analysis we suppose that which is sought to be already done, and we inquire from what it results, and again what is the antecedent of the latter, until we on our backward way light upon something already known and being first in order. And we call such a method analysis, as being a solution backwards. In synthesis, on the other hand, we suppose that which was reached last in analysis to be already done, and arranging in their natural order as consequents the former antecedents and linking them one with another, we in the end arrive at the construction of the thing sought. And this we call synthesis. (Hintikka and Remes, 1974, pp. 8–9; their interpolation.)

The passage continues with a discussion of two kinds of analysis, theoretical and problematical. We are concerned here with the theoretical kind in which

> we suppose the thing sought as being and as being true, and then we pass through its concomitants /consequences/ in order, as though they were true and existent by hypothesis, to something admitted; then, if that which is admitted be true, the thing sought is true, too, and the proof will be the reverse of analysis. But if we come upon something false to admit, the thing sought will be false, too. (*Ibid.*)

There are a number of questions about just what Pappus intended and there is considerable literature on this subject. The question of most interest here is whether in his first sentence Pappus intended passage from what is sought through its logical consequences or something weaker. Hintikka (*ibid.*) argues that the crucial term used in the first sentence means 'concomitants' and is therefore weaker, in which case there would appear to be a difficulty about the falsification suggested in the last line quoted above. But whatever the intention of Pappus – or any who may have tampered with his text[26] – in Commandinus' translation the crucial term in the first sentence is *consequunter* (1588, folio 157).

Thus, whether from Pappus or Euclid, the mathematicians of the late sixteenth century received a definition of the mathematical method of analysis (rendered by translators as *resolutio*) which had the appearance of a more rigorous procedure than the usual versions of the medieval resolutive method, generally described in terms similar to those of the second definition of Pappus.[27] This no doubt influenced Galileo and others in forming their conception of the Greek mathematical method and adapting it to physical arguments. But in this case, contrary to Randall and others, the result would hardly be either the modern method of hypothesis, deduction and verification or that of Zabarella (Randall, 1961), but a quite different method, and indeed this is what we find.

Let us first look at two commentaries on method which are often cited to demonstrate Galileo's debt to the double method of resolution and composition as developed by Jacopo Zabarella. In the first of these commentaries, the resolutive method as it is used by the Aristotelians is explicitly criticized and in both cases the method prescribed is that of the first definition in Pappus. The first commentary is in a letter dated 1616, supposedly composed by Benedetto Castelli, but probably either written or edited by Galileo himself.[28] The Aristotelians are criticized because instead of proceeding deductively (*deducendo*) step by step, they form from their fantasia a proposition from which they go immediately to the conclusion they wish to prove. This might well be the way in which Galileo viewed the method of Zabarella, particularly if, as Edwards suggests, Aristotle's discovery of prime matter is one of the chief examples given by Zabarella (Edwards, 1960, p. 262). In any event, it is in opposition to such a method that we find, in the letter of 1616, insistence on a method in which the conclusion is taken as true, after which one proceeds from it, "deducing first this, then that, then another consequence, until arriving at something true in itself or which has been demonstrated" (*Opere* IV, p. 521). This surely echoes the first definition of Pappus.[29]

The same is true of Galileo's last explicit reference to the resolutive method which is found in the *Dialogue* (to my knowledge he actually mentions this method a total of three times, if we include the reference in the letter of 1616, and it is only in that letter that the compositive method is mentioned). Here, Aristotle's method is explicitly interpreted as that employed by the mathematicians. Simplicio has claimed that Aristotle always began a priori, showing the necessity of his conclusions "by means of natural, evident, and clear principles," and only afterwards supported these a posteriori by means of the senses and earlier tradition (Drake, 1967, p. 50). Salviati responds that this was how Aristotle presented his doctrine, but it was not his method of investigation. For he began with

> the senses, experiments, and observations, to assure himself as much as possible of his conclusions. Afterwards he sought means to demonstrate them. This is what is done for the most part in the demonstrative sciences; and this comes about because when the conclusion is true, one may by making use of the resolutive method encounter (*si incontra*) some proposition which is already demonstrated, or arrive at some known principle (*per se noto*); but if the conclusion is false, one can go on forever without finding any known truth – if indeed one does not encounter some impossibility or manifest absurdity. (*Ibid.*, p. 51; *Opere* VII, p. 75.)

Again Galileo clearly follows the first definition of Pappus, for the method may lead to a false or absurd proposition and this would only happen through rigorous deduction of consequences from what is to be proved, in the event that it is false. Also, as is evident, it is the way of proof which is to be discovered, not previously unknown principles which will now be made known through analysis and synthesis. This again points to the method of the mathematicians rather than the scholastics, and this interpretation is further confirmed by the example used for illustration – again, the Pythagorean theorem.

It appears then that Galileo, in the old *De motu*, followed either a scholastic definition of the resolutive method or the second definition of Pappus, or perhaps some mixed tradition since he is searching for a *cause* which is an established principle. By time of the *Dialogue* (or earlier, in view of the 1616 letter), he has adopted the first definition of Pappus. Now, what can be the significance of this change from the second definition to the first? In practice it seems to be virtually nil. For in actual practice Galileo seems to follow the second definition as much as, or more than, the first. (See Wisan, 1974, section 6, for examination of the method of analysis in Galileo's mathematics, and below for a physical argument which seems roughly to follow such a method.) Why, then, the later emphasis on the procedure laid down in the first definition? The answer will emerge, I believe, if we keep in mind Galileo's remarks quoted above, concerning the difference between the human and the divine intellect, while examining a purely physical argument in the *Dialogue*. Here, we shall find Salviati actually *discovering* a 'true cause' by a method which resembles that of analysis (or resolution) but is not explicitly so-called. And, although the argument is entirely empirical, both Salviati and Sagredo speak of it as though it employed rigorous mathematical reasoning.

The three participants in the Dialogue are discussing Gilbert's book on magnetism. Salviati begins with high praise for Gilbert's work, but then criticizes him for not being more of a mathematician, remarking that "a thorough grounding in geometry. . . would have rendered him less rash about accepting as rigorous proofs those reasons which he puts forward as *verae causae* for the correct conclusions he himself had observed. His reasons, candidly speaking, are not rigorous, and lack that force which must unquestionably be present in those adduced as necessary and eternal scientific conclusions" (Drake, 1967, p. 406). Salviati then gives his own argument for the true cause of the phenomenon in question, and this argument presumably has that force lacking in Gilbert's. That is, Salviati's argument, unlike Gilbert's, must lead to *necessary* conclusions as in geometry. But Salviati's explanation is not geometrical, it is not quantitative and it is not deductive.

The phenomenon to be explained is the increase in the strength of a lodestone when equipped with an iron armature. Salviati observes (as did Gilbert) that the armatured lodestone does not attract objects through a greater distance but only holds them more strongly, and that this difference disappears if a piece of paper is inserted between the armatured lodestone and the object. Therefore, he concludes, there is no change in the force itself, but only in its effect. Then, "since for a new effect there must be a new cause," he seeks to discover what is changed when objects are supported from the armature instead of the bare lodestone (*ibid*., p. 407). Since where the object originally touched the lodestone, it now touches iron, Salviati says it is necessary to conclude that change in the contact causes the difference in the effect. Thus far, Salviati reasons from the method of difference. Now he must form an hypothesis which will explain why there is a difference in contact. He observes that the 'substance' of the iron is finer, purer and denser than that of the lodestone (*ibid.*). This observation suggests that when the object touches iron there are a greater number of points of contact. Experiments with a needle confirm the hypothesis. According to Salviati, the point of the needle attaches itself no more firmly to the armature than to the bare lodestone, but the larger end of the needle holds more firmly to the armature. From these arguments, Salviati concludes (somewhat rashly, one might say) that the strength of magnetic attraction holding two objects together depends on the number of points of contact.

Salviati then describes another experiment which he once performed and which permitted direct observation of the true cause of the difference in contact. He had a lodestone polished very smoothly until the pure substance of the lodestone could be distinguished from the impurities in it. It could then be clearly seen that iron filings were attracted to those parts of the lodestone which were pure, but not to the innumerable little particles of other stone mixed with it and which appeared as spots of a different color (*ibid.*, p. 409). Thus the difference in the denseness of the iron and that of the lodestone, which was sensibly known (*sensatamente si conosceva*), explained the difference in contact. Salviati could directly see, or touch with the hand as he puts it (*toccai con mano*), that which he sought – the true cause of the phenomenon he was investigating (*ibid. Opere* VII, p. 435).

Now, Salviati's conclusion might be (but is not) stated mathematically: the force of attraction seems to be proportional to the number of points of contact. But no mathematics has been used to establish this result and the argument does not proceed deductively. Salviati has used only the method of difference, together with observation and experiment and the logical law that there must always be a new cause for a new effect. The reasoning, however, although not deductive, has proceeded by small steps where each step seems to follow *necessarily*. And, indeed, only by this assumption can Galileo suppose that his demonstration has that force which he found lacking in Gilbert's argument. In fact, by this interpretation, the procedure begins to resemble that of the geometrical method of analysis. According to Pappus' analysis, however, one proceeds from the conclusion to be proved to a previously proven or evident premise from which the conclusion follows. But the impure nature of the lodestone which is the cause of the difference in contact, which is again the cause of the phenomenon to be explained, is not a proven theorem or a previously known principle. So Salviati uses direct empirical observation of the polished lodestone both to discover the cause and to provide its necessary evidentness. The causal connection is established by inference according to the 'laws of logic,' to which Galileo has occasionally alluded (see above, Section II).

In other words, the previously unknown cause of the magnetic phenomena has been encountered as an empirical fact which is established by more or less immediate experience much like that employed to establish 'true and evident' principles of a demonstrative science. But as in the case of the principle of affinity in the *Discourse on Bodies in Water* (which was also a new discovery), the cause which has been found seems less rationally evident than principles such as those insisted upon in the old *De motu* or in early correspondence concerning the principles for the new science of motion (see Section I), and confirmation requires more extensive arguments from experience. Thus, the general shift we have seen towards a greater degree of empiricism seems to extend beyond astronomy to both mathematical and non-mathematical physics. But, in fact, Galileo's intention may be just the reverse. For the whole thrust of the argument on magnetism is the establishment of a 'true cause' which Galileo, unlike Gilbert, has properly established by using methods in some sense mathematical. Sagredo remarks more than once that he finds Salviati's arguments and experiments "very little lower than mathematical proof" (Drake, 1967, pp. 408, 410). And Salviati describes the method followed in terms suggestive of those used to describe the analytic method of the mathematicians:

> In investigating the unknown causes of our conclusions, one must be lucky enough right from the start to direct one's reasoning along the road to truth. When travelling along that road, it may easily happen that other propositions will be encountered which are recognized as true either through reason or experience. And from the certainty of these, the truth of our own will acquire strength and certainty. (*Ibid.*, pp. 408–409.)

From these various remarks linking the empirical argument on magnetism with mathematical reasoning, we must conclude that proceeding step by step in accordance with the 'laws of logic' for inferring causes from effects, combined with direct observation of the cause, is tantamount to or 'very little lower than' mathematical proof. It would appear then that it is a higher order of rigor that Galileo sees in the first definition of Pappus, whereas the less rigorous method of the scholastics would be suggested by the second. Thus, he describes the analytic (resolutive) method in terms of the first primarily because it suggests the rigor and precision of the mathematical tradition as opposed to that of the scholastics. His choice of definitions provides a clue not to actual logical procedures but rather to the ideal and source of inspiration found in the works of the mathematicians. This distinction may seem overly subtle, but I believe it is an important one. Above all, what Galileo sees in arguments by mathematicians as opposed to those by Aristotle and his medieval followers is that the former proceed only by *small* and *necessary* steps from clearly defined terms and assumptions which are both explicit and evident. These are the essential features of mathematical arguments and what gives mathematics its certainty. And Galileo evidently believes one can use 'logical laws' for inferring causes from effects with very nearly the same degree of rigor.

In fact, it is just such laws of logic which are employed by Galileo to establish that the earth's motion is the 'true and primary' cause of the tides. This is done in the fourth day of the *Dialogue* where Salviati drops his mask and argues in a straightforward manner for the motion of the earth. There is no further reference to Plato's theory of knowledge, and the discussion no longer proceeds by Socratic questioning. The key argument employs the method of concomitant variations as a rule of inference from which it follows that the earth's rotation is the cause of the tides.

> Thus I say that if it is true that one effect can have only one basic cause, and if between the cause and the effect there is a fixed and constant connection, then whenever a fixed and constant alternation is seen in the effect, there must be a fixed and constant variation in the cause. . . .Hence it is *necessary* that whatever the primary cause of the tides is, it should increase or diminish its force at the specific times mentioned. . . [and therefore] . . . the primary cause of the uneven motion of the vessels, and hence of that of the tides, consists in the additions and subtractions which the diurnal whirling makes with respect to the annual motion." (*Ibid.*, pp. 445–446; my italics.)

Galileo employs geometry to demonstrate (fallaciously) variations in the speeds of places on the earth depending on the daily, monthly and annual motions. But he does not, as in his arguments for the contiguity of sunspots, rigorously deduce from his hypotheses conclusions which are verified through observations. He argues qualitatively rather than quantitatively that the daily, monthly and annual motions of the earth-moon system must produce exactly the increases and decreases required by the method of concomitant variations. Thus, again, the argument is not strictly mathematical, and it is reinforced by

Galileo's claim to have performed an experiment showing that tides could not occur on a stationary earth (thus eliminating most alternative hypotheses). But Sagredo feels that he has been led by sure and easy steps to a most sublime conclusion:

> You, Salviati, have guided me step by step so gently that I am astonished to find I have arrived with so little effort at a height which I believed impossible to attain... just as climbing step by step is no trouble, so one by one your propositions appeared so clear to me, little or nothing new being added, that I thought little or nothing was being gained. (*Ibid.*, p. 454.)

Again, a non-deductive argument is analyzed in terms suggesting a strictly deductive proof where 'little or nothing new' is added at each small step. Sagredo's words are no doubt intended to echo the earlier discussion of mathematical reasoning, while the argument on magnetism at the end of the third day may have been intended to illustrate the quasi-mathematical force of the 'laws of logic' for inferring causes from effects. In the case of the earth's motion, of course, there is no possibility of direct observation of the cause. However, according to Sagredo, Salviati's experiment which is supposed to show that the observed tides could never occur if the seas were stationary seems "to admit of no refutation" (*ibid.*, p. 461). Simplicio, apparently assenting to the force of Salviati's arguments, takes his final stand on the doctrine that God is not constrained by man's logic and that the tides may be caused on a mobile earth by a miracle. This argument, of course, has already been discounted and Salviati allows it to stand. He has made his point.

No doubt the sheer complexity of the data to be explained assured Galileo that no hypothesis other than the earth's motion could account adequately for the observed phenomena. And although his rules of inference from effects to causes had to be supplemented by direct observations or other arguments, he clearly saw them as very powerful rules, nearly equivalent to the process of mathematical deduction, or the process of reasoning step by step from one clear, distinct and necessary idea to another. Such ideas may be drawn from experiments and observations as well as by deduction in the geometrical sense. Thus, Galileo appears to be approaching Newton's 'deductions' from the phenomena. However, as we shall see in Section V, Galileo could not – or would not – use this procedure to ground his mathematical treatise on motion.

V. REVERSION TO RATIONALISM? THE TWO NEW SCIENCES

Galileo's thinly disguised defense of Copernicus and criticism of the Pope brought on his subsequent trial and confinement and very nearly put an end to his scientific research. Yet, even as he began his last years under house arrest, Galileo was putting together the materials that make up the discourse on *Two New Sciences.*[30] These discourses are again in dialogue form and the same three friends carry on their discussions as before. The third and fourth days include the new science of motion and the second presents a new science dealing with the strength of materials. The first day is devoted to a long informal conversation covering such topics as the nature of matter and resistance of solid materials to separation and breakage, the way bodies fall through different media, pendular motions and musical harmony. It is here that we find Galileo's most mature remarks on the methods of discovery and proof in science, and this material, together with the formal treatises in the second, third and fourth days, probably provides the most complete view possible of Galileo's final position on science and method.

We find him still assuming the mathematical model: true conclusions must be derived from true and evident principles. There is more emphasis on experience than in the early *De motu,*[31] and principles are investigated and established by a greater variety of arguments, including some which are similar to those used earlier for inferring causes from effects. But the truth of principles must still be shown in as direct a manner as possible. Indeed, it appears that there must be an *exact* correspondence between fundamental mathematical propositions and visual demonstrations of such propositions.[32] Deductions from these propositions are then regarded as true without further empirical confirmation.

This was the ideal. But Galileo's science of motion (in the third and fourth days) required principles which he could not establish according to this ideal. And despite the tendency to merge mathematical and empirical arguments that we saw in the *Dialogue*, he hesitated to ground his new science by the same kind of arguments used, for example, in the discussion of magnetism or the establishment of the hypotheses of astronomy. Consequently, he had deep difficulties in establishing a foundation for the treatise on motion. Exactly what the problem was can best be understood if we first look at the way in which principles are discussed in the first day of the *Two New Sciences*.

The first topic of the *Two New Sciences* is the problem of resistance of matter. Salviati tells us that our 'Academician' has "demonstrated everything by geometrical methods so that one might fairly call this a new science" (Crew and de Salvio, 1914, p. 6). Admittedly, some of his conclusions were already known and Aristotle's writings on this subject are mentioned. But Salviati remarks that the results already known "are not the most beautiful and, what is more important, they had not been proved by necessary demonstrations from primary and indubitable foundations" (*ibid.*). Galileo, then, has established a science dealing with the strength of materials or, in his terms, resistance to separation and breakage, and this science consists in mathematical demonstrations from indubitable first principles.[33]

Before development of the new science there is a long digression in search of the cause of the phenomena in question, that is, the cause of cohesion, which is sought in the structure of matter. At the beginning of this search Salviati remarks that they will adhere to no narrow, restrictive method, but by discoursing freely will uncover new truths (perhaps a commentary on the Renaissance obsession with method?). And, indeed, arguments are drawn from mathematics, from experience and from the 'laws of logic.' But no satisfactory conclusion is found. Salviati sets forth an elegant mathematical model for the structure of matter and the transformation of solids into liquids, but the model is supported only by analogies, not by a visual instance of the model itself. Thus, Salviati does not claim to have established his conclusion or to have found the underlying cause of cohesion.[34] In the second day the science concerning resistance is formally developed on the basis of principles from mechanics without reference to the cause of the phenomenon being studied, just as in the third day accelerated motion will be treated independently of any supposed cause of that motion.

The foundation of the science of resistance seems to be unproblematic. A deeper causal explanation is lacking, but the science can still be satisfactorily developed on the basis of known and evident principles. This is not the case with the science of motion. Galileo's difficulties with the principles of this science can be better understood if we examine a discussion in the first day which establishes an important proposition on motion. It is in the context of this discussion that we find some of Galileo's clearest statements of the criteria for establishment of principles.

The proposition, which is referred to as a principle and which does in fact function as such (Crew and de Salvio, 1914, p. 75) is that all freely falling bodies fall through equal distances in equal times, or as Salviati puts it, with equal speeds. This, of course, means that all bodies have the same acceleration near the surface of the earth, but Galileo was seldom clear in distinguishing between acceleration and speed. I shall, therefore, refer to this as the principle of equal times in order to distinguish it from the postulate that bodies descending along inclined planes acquire equal speeds in descent through equal vertical distances (which I refer to as the postulate of equal speeds).

Galileo's conception of what must be done to establish his principle is clear from Salviati's remark that

> this idea is, I say, so new, and at first glance so remote from fact, that if we do not have the means of making it just as clear as sunlight, it had better not be mentioned; but having once allowed it to pass my lips I must neglect no experiment or argument *[esperienze ou ragione]* to establish it. (*Ibid.*, p. 83; *Opere* VIII, p. 127).

This language is itself clear as sunlight: principles must be made clearly evident through experience and reason, and, to be sure, the principle of equal times is established by numerous arguments, including a more or less direct experiment showing that the periods of pendula of equal length are very nearly equal for bobs of different densities. (See Naylor, 1974b, 1976b for detailed examinations of Galileo's experiments with pendula.)

Afterwards, Sagredo praises the 'Academician' (Galileo, of course) for his clear explanation

> which like everything else of his is extremely lucid, so lucid, indeed, that when he solves questions which are difficult not merely in appearance, but in reality and in fact, he does so with reasons, observations and experiments *[esperienze]* which are common *[tritissime]* and familiar to everyone. (Crew and de Salvio, 1914, p. 87; *Opere* VIII, p. 131.)

Then, again, sounds the *leitmotif* we have heard through all of Galileo's discussions of demonstrative science: Sagredo adds that it is a "most admirable and praiseworthy feature of demonstrative science, that it springs from and grows out of principles well known, understood and conceded by all" (*ibid.*).

In fact, the proposition – or principle – of equal times has not, in the traditional sense, been derived from deeper principles. Like the proposition on magnetism (see Section IV, above) it has been established by numerous arguments from experience and experiments, including an experiment which seems to provide a direct confirmation (hence, perhaps, the ambiguity about whether it is a principle or a proposition derived from other principles). But the discussion of this proposition shows that these are still Galileo's criteria: principles must be manifestly true, well-known, and accepted by all.

Let us now turn to the science of motion in the third and fourth days of the *Two New Sciences* and examine the problem of its foundation. In the first place the principles of the new science were themselves new and not well known, much less widely accepted. Morever, they could not be rendered manifestly evident through exact visual demonstration. In fact, Galileo himself was long in coming to a secure grasp of the principles underlying his new science. He vacillated between two different concepts of uniformly accelerated motion (was the acquired velocity proportional to distance or to time?) until well after his move to Florence in 1610, and he seems for a time to have confused the incorrect concept with his postulate of equal speeds (Wisan, 1974, esp. section 5.4). Then after choosing the correct law of fall he still had no satisfactory empirical demonstration of this law or of the postulate. To confirm the law of fall he had the experimental demonstration of one of its consequences, to wit, the times-squared law. To confirm the postulate of equal speeds he had a demonstration with a pendulum from which the motion of bodies along inclined planes could be inferred by analogy. But neither of these provided a direct and exact visual demonstration of the principle to be established and, in fact, Galileo could only confirm theorems derived from these principles.

It is perhaps for this reason that in the third day of the *Two New Sciences* we find Galileo using language in a manner that sometimes suggests the method of hypothesis, deduction and experimental confirmation. He begins by drawing attention to Archimedes' treatise on spirals, although Archimedes is not mentioned by name. According to Galileo, 'some' have invented motions (including spiral motions) not occurring in nature and from definitions of these motions have demonstrated propositions *ex suppositione* (Drake, 1974a, p. 153; *Opere* VIII, p. 197). Galileo's book on naturally accelerated motion, however, is to be based on definition of a motion which does occur in nature, and he claims that results of natural experiments (*naturalia experimenta*) are in agreement with properties of motion demonstrated from his definition.

The experiments, however, are not described until later, and the formal statement of the definition includes no reference to natural motion. It simply states that

that motion is equably or uniformly accelerated which, abandoning rest, adds on to itself equal momenta of swiftness in equal times. (*Ibid.*, p. 154.)

Like Archimedes' definition of spiral motion, Galileo's *formal definition* does not assert existence of the motion defined. Sagredo then remarks that since all definitions are arbitrary it would be unreasonable for him to oppose this one. However, he does doubt whether the definition

conceived and assumed in the abstract, is adapted to, suitable for, and verified in the kind of accelerated motion that heavy bodies in fact employ in falling naturally. (*Ibid.*, p. 155.)

Salviati replies by first giving a rational argument from the simplicity of nature, after which he gives additional arguments from experience which support continuity of motion but not any particular formula. Then, as in astronomy, he adds a rational refutation of the most likely alternative to the definition he has given.[35]

At this point the *definition* is said to be established and the postulate of equal speeds is introduced as the *solo principio* on which the treatise is based (Drake, 1974a, p. 162; *Opere* VIII, p. 205). Now, whatever the definition did, or did not, require, the postulate – a *principle* – must be established by rendering it evident through exact visual demonstration. Unfortunately, this is not possible, according to Salviati (*ibid.*, p. 164; p. 207). The best he can do is describe an experiment with a pendulum which is very nearly a 'necessary' demonstration (*ibid.*, p. 162; p. 206). The pendulum, consisting in a bob on the end of a string, is suspended from a peg on an upright board. A nail is placed on the board along the vertical from the point of suspension so that when the bob reaches its lowest point the motion of the string will be impeded by the nail. Salviati argues that wherever the nail is placed (provided, of course, it is not too near the bob as the bob passes beneath it) the bob will rise again to the same height from which it descended. But it will rise along different paths depending upon where the nail is placed: the higher the nail the longer the path. The point, of course, is that the bob acquires the *momento* needed to carry it back to the same height regardless of the path it takes. Salviati claims, then, that the bob must acquire equal *momento* (and therefore equal speeds) in descending from equal heights regardless of path, and he asserts that an ideal object on an inclined plane would behave in an analogous manner.

Sagredo, who is already disposed to accept the postulate by force of the *lume naturale*, greatly admires the demonstration and would accept it as a necessary proof (*ibid*., pp. 162, 164; pp. 205, 207). Salviati, however, admits that the experiment cannot be regarded as a necessary demonstration of the postulate because acceleration along the arc of a circle does not proceed *exactly* as it does along an inclined plane. There is, in other words, a *theoretical* difference between the two cases so that the experiment, however precise in its results, cannot give an exact visual demonstration of the principle. But, he adds, the *absolute truth* of this postulate will be established when it is shown that conclusions drawn from this "hypothesis correspond *punctualmente* with experience" (*ibid*., p. 164; p. 208).

This last statement implies that the postulate is to be established by the method of hypothesis, deduction and experimental verification as this is understood today. Yet, not a single experiment is offered to justify the postulate in this way! Nor does he appear to have looked for such an experiment (see Wisan, 1974, p. 124). Instead, he continues to search for a more immediate demonstration or a deeper principle from which the postulate might be derived (*ibid*., section 5.5). Later, in a letter to Baliani, August 1, 1639, replying to criticism of the way in which he tried to establish his postulate, Galileo admits that the principle he has supposed does not seem to have that *evidenza* for which one looks in principles which are to be assumed as known (*principii da supporsi come noti*) (*Opere* XVII, p. 78). In this letter Galileo promises to supply such evidence, but the 'postulate' is eventually made to depend on the definition of uniformly accelerated motion and another principle smuggled in from 'mechanics' (Drake, 1974a, pp. 174–175).

Galileo has not deviated from his initial assumptions in the old *De motu* concerning the principles of a demonstrative science: such principles must be made immediately evident. But he has not fully succeeded in doing this for the postulate, and his failure is even more striking in the case of the principle that velocities acquired in free fall are proportional to time of fall. It was too difficult to demonstrate this directly in even an approximate manner, and, as we shall see, it is probably for this reason that Galileo treats this principle formally as an arbitrary definition. At the same time, however, he does also claim to show indirectly that the motion he has defined does exist in nature.

After deriving the times-squared theorem from the definition (and the first theorem on accelerated motion, which is based on that definition and additional but unstated assumptions), Salviati describes the famous inclined plane experiment which directly confirms the times-squared theorem and, of course, indirectly confirms the definition. The language used, however, suggests a more or less direct confirmation of the definition, and Salviati refers to this procedure as *customary* in such sciences as perspective, astronomy (!), mechanics, music, and "others which confirm with *sensate esperienze* their principles, which are the foundation of the subsequent structure" (*ibid*., p. 169; *Opere* VIII, p. 212). In other words, such experiments (or experiences) confirm the principles underlying the subsequent structure, which structure needs no further confirmation. There is still no hint that Galileo would regard confirmation of remote consequences as relevant to establishment of principles in a mathematical science. Yet, the fact is that his experiment *only* confirmed consequences; but this was the best he could do and so he tries to slide by this point. Significantly, no more experiments are offered to confirm consequences following from the foundation.[36]

The method of the *Two New Sciences* is clearly not that of hypothesis, deduction and experiment in the modern sense. In fact, Galileo was quite unable to treat the principles of a demonstrative science as hypothetical for they must be true and evident. They are not merely rational for they must be confirmed by experience. This, however, does not mean confirmation of consequences, but rather the process by which the principles are to be rendered evident to the intellect. Unfortunately, the new science required principles which were *not* immediately evident in the sense required and could not be made so.

Galileo struggled with this vexing problem for many years but never found a satisfactory solution. It may be, however, that the establishment of his principles was not, for him, the matter of greatest importance. For, from the time-squared theorem and the postulate, both at least roughly confirmed by direct experience, Galileo had derived enough results on accelerated motion to make up a whole 'book' on this topic, and two additional propositions, also supported by reason and experience, gave him his book on projectile motion.[37] Together, with a brief, initial treatment of uniform motion, these made up a substantial mathematical treatise and it is the treatise as a whole, rather than the presentation of new principles, that Galileo seems to have regarded as his greatest achievement. For him, as for Aristotle, there is no scientific demonstration of principles. It is the body of demonstrated propositions that constitutes the science. This was the glory of the Greek mathematicians who were famed, not so much for their fundamental principles as for their elegant mathematical demonstrations. Similarly, Galileo saw the chief merit of his own work, not in his principles but in his deductions from those principles. It was his mathematical propositions which were his important discoveries. Galileo was jealous of his claims to many inventions and discoveries, but he was quite reticent about his discovery of new *principles*. And, indeed, this is just the attitude we should expect if we suppose that the principles of a demonstrative science should ideally be already known and accepted.

Therefore, whether or not his principles were acceptable, Galileo felt that he had a treatise on motion which should rank with the works of the Greek mathematicians. This, no doubt, is why he introduced his book on accelerated motion by reference to Archimedes' treatise on spirals. Just as Archimedes arbitrarily defined spiral motion and derived mathematical propositions from this definition, so Galileo arbitrarily defined a uniformly accelerated motion and proceeded in the same way. If Galileo's definition should be disputed, he has nonetheless done exactly what Archimedes did, or so he thinks (Wisan, 1974, section 1.2). Thus, he has Sagredo first emphasize that in mathematics definitions are arbitrary and later point out that results rigorously demonstrated from such definitions (*ex suppositione*) must be accepted as conclusive (Drake, 1974a, pp. 154–155, 222).

This line of reasoning is further elaborated in correspondence of 1637 and 1639, where Galileo remarks that his propositions are derived *ex suppositione* and that even if the motions he supposed did not exist in nature, his demonstrations "founded on my supposition, lose nothing of their force and conclusiveness, just as nothing prejudices the conclusions demonstrated by Archimedes concerning the spiral that no moving body is found in nature that moves spirally in this way."[38]

Galileo then argues that the motion he has supposed (unlike the spiral motion of Archimedes) *does* exist in nature as his experiments show. Here the Florentine mathematician seems to be suggesting that he has gone beyond the Greeks in producing a rigorous mathematical treatise on the subject of physically real motions. But he also cautiously presents his treatise as purely mathematical, and he may well have been uncertain which claim was to be preferred. He was aware that his principles were not immediately evident and only indirectly established, and he perhaps expected (as he suggests in the case of the science of resistance) that more fundamental principles might yet be revealed. Yet the treatise was clearly more than a work in pure mathemetics. From the times-squared law and the postulate, both of which he was convinced were physically true, Galileo had been able to make many further discoveries concerning the 'properties' of accelerated motion. Indeed, Galileo's greatest innovation was perhaps as he himself saw it: the use of Greek geometry not only to represent the path of naturally moving bodies but to show how such bodies would move along different paths in different times, given different initial conditions (Wisan, 1974, section 6). And, if granted his law of fall (the definition of uniformly accelerated motion) and the double-distance rule derived from that law, he could solve such problems as determining the 'impetus' of a projectile at any point on its path and construct tables supposedly useful for artillery (Wisan, 1974, section 7).

Galileo was quite aware that he was not the first to apply mathematics to *nature* (something modern historians sometimes forget) but he knew he had opened the door to a new way of studying motion and he rightly foresaw that further research along this line would be of immense interest. Indeed, he seems to have foreseen, however dimly, the future direction of rational mechanics. His treatise with its clumsy geometrical methods and lack of dynamic principles scarcely resembles the works of the eighteenth-century analysts. But in a limited way he was trying to do much the same thing. At the very least, his treatise was the herald of a new form of discourse on precisely that subject which was most crucial in the development of rational mechanics in the eighteenth century.

VI. CONCLUSION

Galileo is justly celebrated for his experimental and observational approach to science and for many philosophically significant remarks and arguments which reinforced, perhaps crucially, the movement towards 'mathematization of nature' and the acceptance of the Copernican system. Also, of course, he provided certain key propositions (above all, the times-squared theorem) which played an enormously important role in the development of Newton's great synthesis of terrestrial and celestial mechanics.

My examination of certain regressive looking peculiarities in Galileo's conception of scientific method, particularly in his science of motion, may seem merely to darken this generally accurate picture to no purpose. But there still remains much to be understood concerning precisely what brought about the radical change between medieval and modern science in the west and just what the most significant steps were. Galileo is obviously a key figure and his evident contributions include both empirical and rational elements. But one aspect of his work that has been much neglected is that which most resembles the rational mechanics of the eighteenth century from which came much of the paraphernalia essential to the extraordinary mathematical technology of today.[39]

The study of Galileo's work on motion may not unlock the whole secret of how modern mathematical science started nor take us very far into its development. But his labor of some fifty years to bring forth a mathematical treatise on motion may be an important key. Although one seldom associates Galileo's rather primitive work with rational mechanics after Newton, who is legitimately regarded the real founder of modern mechanics, it was Galileo who transformed the traditional physics of motion into a geometrical mechanics in the Archimedean-Jordanian tradition and initiated exploration of mathematical consequences of principles shown in some more or less direct way to be true of natural, terrestrial motions.[40]

How important this step was for future developments is difficult to determine. Precisely how Galileo's treatise influenced whom will no doubt remain a matter of speculation. But we should at least examine closely just what procedures are new in Galileo's treatise and available for the use and inspiration of those who came afterwards. These were, I believe, essentially two: first a mathematical technique for discovering, from certain initial conditions, "what events take place," as Galileo would put it (Drake, 1974a, p. 197, *Opere* VIII, p. 243), when a body is moving near the surface of the earth; second, the systematic elaboration of a substantial body of theorems which showed very precisely how such techniques might be applied in future studies of motion.[41] This, surely, is what Galileo had in mind when he claimed to have created a new science, and it seems reasonable to say he was right.

Brooklyn College

NOTES

* I would like to mention here the great benefits I gained from the conference out of which this book has come (Blacksburg, Virginia, October, 1975). There was much vigorous and fruitful discussion and exchange of ideas has continued. Also, I wish to thank Peter Machamer and Charles B. Schmitt for helpful comments at later stages of my paper.

[1] Speaking very generally, I believe it is fair to say that the focus of interest in Galileo's method tends to shift back and forth from his empiricism and experimentation to his abstractions from physical reality and use of mathematics. A widely accepted thesis in the nineteenth century was that Galileo destroyed abstract Aristotelian science by inventing modern experimental method. This traditional empiricist interpretation came under attack early in this century until it was virtually reversed by A. Koyré, who characterized the scientific revolution in general, and the work of Galileo in particular, as the victory of Platonic rationalism over Aristotelian empiricism (1939). Koyré's more tempered rationalism or a modified empiricism. On the whole, however, emphasis in the last few decades has been on Galileo's mathematical as opposed to his experimental method. But now, with discovery of evidence in Galileo's manuscript notes showing that he did, in fact, perform some experiments (Drake, 1973, 1975b), and the emergence of new evidence linking Galileo with the scholastic tradition (Wallace, 1974, Crombie, 1975, paper by Wallace in this volume), there will doubtless be a general resurgence of interest in Galileo's empiricism (see, for example, Naylor, 1974a,b,c, 1976a,b).

[2] To mention just a few more results at hand or in progress, P. L. Rose has been tracing manuscripts relating to major figures and has published an extensive study of the mathematical renaissance in Italy (1976). Rose and Drake have sketched the fortunes of the *Mechanics* of Pseudo-Aristotle (1971), and Drake and Drabkin have provided partial translations of several mechanical treatises of the sixteenth century (1969). P. Galluzzi has an important paper on writings by Neo-Platonists who were known to Galileo (1973), and G. Giacobbe has examined the question of mathematical certitude in Piccolomini and Barozzi (1972a, 1972b). L. Sosio's study of Paolo Sarpi's *pensieri* on motion (1971) shows how a topic of deep interest to Galileo was viewed by a close friend and suggests ways in which Galileo's circle of friends in Venice and Padua may have contributed to the shaping of his thought. An edition of Sarpi's scientific writings, now being prepared by L. Sosio and L. Cozzi will soon provide a rich source of information concerning other scientific and philosophical ideas to which Galileo was very likely exposed. T. Settle is exploring Galileo's Florentine background with particular attention to the Academia del Disegno and Ostilio Ricci (1968). Also, hopefully, we shall soon have access to a long awaited volume by A. C. Crombie and A. Carugo which will reveal much new material on Galileo's sources (see Crombie, 1975). Meanwhile, further research is being done in this area by W. A. Wallace, C. B. Schmitt and others.

[3] Today, of course, everyone knows that one cannot argue rigorously from effects to causes, and it is customary to say that Mill's methods were of no importance in the development of science and that Bacon was altogether without scientific influence. However, Galileo's use of these methods as rules of inference (see Section IV) is sufficient evidence to show that they had at least *this* importance in the development of science. Whether Bacon himself exerted any influence on Galileo is a more dubious matter. For the method of difference Galileo cites Buonamico's treatise on motion which was

published in 1591 (see Crombie, 1975). However, I find no evidence that Galileo employed, or even thought about these rules before 1611 (see Section II), and I do not find the method of concomitant variations in his writings before the *Dialogue*. This suggests the possibility that, although the methods were long known (even suggested by Aristotle), Bacon may have had some influence after all, at least as a publicist.

[4] In *Opere* I, pp. 251–419. The main essay and some of the remaining material has been translated in Drabkin and Drake (1960). For analyses of this work see Koyré (1939), Moody (1951), Fredette (1969, 1972), Drake (1976). See also Drake and Drabkin (1969) for translation of some additional material.

[4a] Galileo does not, of course, define 'matter,' 'density' or 'relative density' in their precise modern senses. Just what he meant by 'matter' and whether he ever distinguished between weight and mass are interesting questions but not relevant to the present discussion. 'Matter' can therefore be read here in an intuitive way, and where Galileo's concept is that denser objects contain more matter in a given volume than do less dense objects, I will use the terms 'density' and 'relative density' to render his meanings.

[5] For Galileo's attempts to improve on Archimedes, see Section II below and Shea (1972, Ch. 2). On the Euclidean theory of proportion, see *Opere* VIII, pp. 349–362 (the so-called 'fifth day' of the *Two New Sciences*).

[6] It is defined and used in the first five propositions of Book XIII of Euclid's *Elements*, but generally believed to be a somewhat later interpolation (Heath, 1956, Vol. III, p. 442). Proclus, well-known, of course, in Galileo's day, commented on the method of resolution and composition (1970), and it appears with ecstatic commentary by Dee in Billingsley's translation of Euclid (1570, fol. 397v). It also appears in some other translations of Euclid (notably that of Commandino) and it was certainly known to Vieta (Klein, 1968, p. 154). However, the fuller elaboration by Pappus may have been important, especially after this became available in Latin. Although Galileo rarely used the terms associated with the method of resolution and composition – so far, I have found three instances of 'resolution' in this sense, one of which is in a letter supposedly by a student, and 'composition' appears only with the latter (see Section IV) – a considerable amount of legendary material surrounds Galileo's use of the method. I have not traced this to its source; however, it goes back to the early nineteenth century, if not before (see references in Höffding, 1955, pp. 176, 509n). Cassirer links Galileo with Zabarella (1922), and Randall's very influential paper on the subject (1961) has firmly established the legend in our time. Randall's thesis, however, has been critized (especially by N. Gilbert, 1963), and further questions about a link between Galileo and Zabarella are raised by the results of C. Schmitt's research (1969). For detailed studies of Zabarella see Edwards (1960) and Poppi (1972). See also Jardine (1976).

[7] Already argued cogently by Settle (1961, 1966, 1967), and now further confirmed by Drake's manuscript discoveries (1973, 1975b).

[8] As Drake shows (1975b), folio 107 suggests that Galileo did have a direct confirmation of the times-squared law since this is equivalent to the odd number law exhibited in the data on that page of the manuscript. Whether the odd number law was confirmed by singing to a ball rolling down an inclined plane (Drake, 1975b) or by some other means of measurement, the figures on filio 107 are convincing as experimental data which closely conform to the mathematical law. For the postulate, Galileo probably already had his argument from the experiment with the pendulum.

[9] According to Zabarella the double method of resolution and composition requires that once the cause has been discovered by resolution and confirmed through further 'mental consideration,' there must then be a return through *demonstratio potissima* to the effects in order that the cause yield distinct knowledge of the effect (Randall, 1961, pp. 57–60).
[10] Salusbury, 1960, pp. 45, 59. The first reference is on p. 45, and the only explanation that is even suggested is on p. 47, where Galileo refers to the altitude which is the "greatest that the Nature of the Water and Air do admit."
[11] (*Ibid.*, p. 59) The experiment described is as follows. Galileo takes a lead ball or bullet (lack of precision about the object used raises the question whether the experiment was actually carried out) and suspends it from one end of a balance. The ball is let down into a container of water while the balance is held in the air. Weight is added to the other end of the balance until the ball is drawn up out of the water. The ball is then beaten into a thin, flat plate and again lowered into the water (still suspended from the balance). Three threads hold it parallel to the surface of the water and again weights are added until the plate is raised from the water. But whereas thirty ounces sufficed to raise the ball out of the water, more weight (unspecified) is required after the ball is beaten into a flat plate. This, of course, is supposed to demonstrate an 'affinity' between the thin plate and the water.
[12] (Drake, 1957, p. 243). Galileo also refers to arguments from experience as "physical logic or *logica naturale* (*ibid.*, p. 268, and Sosio, 1965, p. 233).
[13] Observations of the moons of Jupiter gave him some difficulty due to their speeds and he tells us that a "most scrupulous precisenesse" is required to "calculate their places" (Salusbury, 1960, p. 1). Thus, earlier tables of their motions had to be corrected. Eventually Galileo hit upon a method "which shall serve to bring us to the *perfect knowledge* of the orbits of the moons, and other phenomena as well (*ibid.*, p. 2, my emphasis).
[14] Galileo's letters concerning sunspots were published in Rome in 1613, and are to be found in his *Opere* V, pp. 71–250. They have been partially translated in Drake (1957). I have occasionally altered or extended Drake's translation, in which case I add a citation to the original text.
[15] That a 'clear and unambiguous' mathematical model may provide such a guarantee is sometimes suggested by modern interpretators of Galileo (Clavelin, 1974, esp. p. 445; Shea 1972, particularly in Chapter 7). As will become clear in Section V of this paper, a mathematical model alone does not for Galileo yield truth about the physical world. It is only when a mathematical statement is shown to correspond *punctualmente* with an exact visual demonstration of a physical phenomenon that the mathematical statement yields such truth and thus provides the ground for strict demonstration of further truths. (See note 32 below.)
[16] These remarks occur in Galileo's *Il Saggiatore*, published in 1623. Galileo speaks of *vero accidente, affezzione e qualità*, and *primi e reali accidenti* which are contrasted with *qualità* and *affezzioni*, unqualified (*Opere* VI, pp. 308–310; Sosio, 1965, pp. 261–262; translated in Drake, 1957, pp. 274–475).
[17] At least I see little evidence of such a view and suspect those who give this interpretation of Galileo's philosophy are misreading his remarks about mathematics (see Sections IV and V of this paper for examination of some key passages in Galileo concerning mathematics and human knowledge). There can be no doubt that he finds certainty of mathematics in its logical structure, but it is much less clear that he also finds this

certainty in contemplation of an ultimate reality which is mathematical. The famous remark concerning the 'book of nature' which is written in mathematics is ambiguous in light of the explicit comment that what we can know are mathematical properties as opposed to essences, and nowhere do I find Galileo using the language of the Platonists who associated mathematics with 'true being.' Galluzzi's conclusion that on this point Galileo's philosophical view is closely linked to that of Mazzoni, Barozzi, and others seems unwarranted (1973, pp. 47, 59, 73, 77–79).

[18] See Popkin (1960), esp. Chapter VII. Popkin himself follows the usual view that Galileo was a realist, untouched by the skeptic thought of the time, and indeed Galileo's professions of skepticism may not reflect his real beliefs. But *if* Galileo did *not* hold the belief that ultimate reality is mathematical, then the modified skepticism which I suggest would be quite natural, if not necessary, once Aristotelian essences have to be abandoned.

[19] Naturally Galileo would not want to make too much of this after condemnation of the Copernican system in 1616. However, he does make a number of remarks emphasizing the strength of his arguments for the revolution of Venus about the sun, and in the *Saggiatore*, for example, it is commented that "neither Tycho nor Copernicus could refute Ptolemy for they did not have the evidence from the appearances of Venus and Mars" (*Opere* VI, p. 284). That this crucial evidence was revealed to Galileo and not to Copernicus is explicitly remarked upon in the *Dialogue* (Drake, 1967, p. 335).

[20] Although Galileo has Salviati say that he might have been "much more recalcitrant toward the Copernican system" if it were not for a "superior and better sense" (than that which is natural and common) which joined forces with reason and illuminated him with a "clearer light than usual" (Drake, 1967, p. 328; *Opere* VII, pp. 355–356; Sosio, 1970, p. 393), Galileo never mistakes illumination by such means as a proof. See Section V below, for example, where the *lumen naturale* does not suffice to establish Galileo's postulate that bodies descending along inclined planes of equal heights acquire equal speeds. Sosio interprets the 'superior sense' as a reference to the telescope; however, I do not know of other instances in which Galileo refers to the telescope itself as a 'sense' rather than an instrument and hence doubt this interpretation.

[21] The concession is implicit in Simplicio's response to the argument from the motions of the sunspots: "Unless you first demonstrate to me that such an appearance cannot be accounted for when the sun is made movable and the earth fixed, I shall not change my opinion, nor believe that the sun moves and the earth remains at rest" (Drake, 1967, p. 352). Simplicio explicitly appeals to that logic which teaches him that no hypothesis is rendered necessarily true simply because it saves the phenomena. He has come a long way from his original assumption that there are necessary reasons why the earth cannot move.

[22] Throughout Galileo's writings there are occasional references to a process by which knowledge is 'recalled' or the subject 'awakened,' but the most explicit link with Platonic doctrine occurs in the second day of the *Dialogue* where Simplicio is led to 'recall' that if a stone is cast from the notch of a rapidly whirling stick it will move along a line tangent to the circle described by the stone prior to its release from the stick (Drake, 1967, p. 192). This instance is often cited to show that Galileo held the principle of rectilinear inertial motion and that he established this (and other) principles by Platonic reminiscence. Both these conclusions are unwarranted. It is essential to follow the entire labyrinthic argument beginning where Salviati would persuade Simplicio of the principle of *circular* inertial motion, that is, the principle that since the earth's daily rotation is its

own natural motion then objects on or near it share this motion (*ibid.*, p. 142). Salviati tries to lead Simplicio to this conclusion by Socratic questioning. Simplicio accepts a ship in motion as a model for the moving earth and agrees that a stone dropped from the mast of a ship would behave the same as if it were dropped from a tower on a moving earth. To establish just what would happen, Salviati shifts to the example of an ideal spherical ball on an ideal plane parallel to the surface of the earth. The principle of circular inertial motion is then established by reasoning from undeniable experience that heavy objects spontaneously accelerate downward along a sloping surface but can be made to move upward only through application of force. Consequently, if there is no slope with respect to the earth's surface, the body set in motion will neither accelerate nor decelerate and must move uniformly forever (*ibid.*, p. 148). So what about the stone dropped from the mast of the ship? Seeing the inescapable conclusion, Simplicio simply denies the stone would retain its impetus and argues that forward motion would be impeded by air and also by downward motion. Salviati disposes of these arguments and continues with other examples from experience, some of which entail the Aristotelian assumption that the initial path of an object projected violently is rectilinear. Simplicio accepts this proposition but continues to reject any natural circular motion impressed upon earthly objects and he agrees with the Ptolemaic argument that if the earth rotated all objects on it would be extruded.

Now Galileo believed he could show Ptolemy's argument false without assuming that objects on the earth would participate in a natural rotational motion. That is, even if the earth's motion were a violent rather than natural motion, no extrusion would take place. Therefore, Salviati tells Simplicio that he will refute Ptolemy on grounds of principles (*notizie*) known and believed (*sapute e credute*) by Simplicio as well as himself (Drake, 1967, p. 190; *Opere* VII, 217). The main principle is that the initial path of an object projected violently is rectilinear (Salviati unobtrusively remarks that the projectile would continue to move uniformly and thus, presumably, indefinitely, but this point is not emphasized or developed) and Simplicio is led to 'recall' the Aristotelian principle by Socratic questioning. This process of recollection, referred to as *quoddam reminisci* Drake, 1967, p. 191; *Opere* VII, p. 218), is described as the making of a *fantasia* in the mind, after which the correct conclusion is drawn. Thus the principle appears as an idealized mathematical description of a *fantasia* formed from memory of previous observations. In similar instances the term *concetto* often appears, as when Simplicio is asked what *concetto vi figurate nella mente* (*Opere* VII, p. 48; transl. in Drake, 1967, p. 24), but I do not find *idea* in such a context. Further study of the terms Galileo uses in these contexts and how they are employed by other writers of the time might well furnish deeper insight into his 'Platonism.' See, for example, Panofsky's discussion of *idea* and *concetto*, especially as used by artists in the High Renaissance and by Galileo's contemporaries, particularly his friends in the art world.

[23] Drake, 1967, p. 104, corrected according to *Opere* VII, p. 129 (Drake renders *base comune* as 'equal bases,' which would refer to Euclid I, 36 instead of I, 37, as is intended). Salviati's remark in effect summarizes the proof in Euclid but is not, of course, meant as a formal proof, as is evident both from the context and from the way in which Galileo makes formal geometrical proofs (see Wisan, 1974, esp. section 1.3).

[24] Descartes, 1961, p. 25. See also rules V and XI for further emphasis on proceeding gradually step by step and then trying "to conceive distinctly as many of them /the steps/ as possible at the same time" by running through them in "a continuous and uninter-

rupted process of thought." Similarly, for Galileo the nearer one comes to grasping at once all that is involved in the proof of a theorem the closer one's mind approaches that of God.

[25] There was, however, considerable interest in the logic of mathematics and significant developments soon took place (see, for example, Giacobbe, 1972a, 1972b; Galluzzi, 1973).

[26] Mahoney (1968) tries to account for the apparent logical inconsistency between the two definitions of analysis in Pappus by arguing that the second, which is more commonly found outside of mathematical literature, was actually an interpolation made later.

[27] See Gilbert (1960) and McKeon (1965) for general historical surveys of the methods of resolution and composition (analysis and synthesis). See also Creseini (1965) and Hintikka and Remes (1974).

[28] See *Opere* IV, p. 521, for the passage in question and pp. 13–16 for commentary on the authorship of the *Discourse* in general and the difficulty about this particular passage.

[29] The definition of analysis in Euclid is similar to the first definition of Pappus, and Galileo may have been following Euclid rather than Pappus. However, his writings reveal no sign of this approach to the analytic method until after he surely knew Pappus, first cited in Galileo's early mechanical writings of around 1600 (see Drabkin and Drake, 1960, p. 172). In particular, some early notes on Archimedes' work *On the Sphere and the Cylinder*, where Archimedes explicitly refers to the methods of analysis and synthesis, include no comment on these methods (*Opere* I, pp. 229–242).

[30] See *Opere* VIII, pp. 190–313. The standard English translation has long been that of Crew and de Salvio (1914); however, an excellent new translation by Drake (1974a) is now available. Although Drake's translation is generally superior, particularly for the third and fourth days, the older translation is sometimes preferable for the first two days. As usual, I will indicate the translation most closely followed and wherever I have made alterations the original text will also be cited.

[31] It is interesting to note that in the old *De Motu* Galileo remarks at one point that "when I could not discover a reason for such an effect, I finally had recourse to experiment. . ." (Drabkin and Drake, 1960, p. 83). In the *Two New Sciences*, on the other hand, we find him saying, "But where the senses fail us reason must step in" (Crew and de Salvio, 1914, p. 60). This last remark might well be read in connection with a passage in the *Discourse on Bodies in Water* (Salusbury, 1960, p. 42), where Galileo speaks of an internal contemplation of the nature of water and other fluids. In these last two passages the topic is the same. In neither case does reason (or contemplation) lead to certain knowledge.

[32] This notion of 'exactness' must be qualified to include cases in which negligible variations may occur due to physical impediments. But variations, however negligible, if due to a *theoretical* difference cannot be allowed. See, for example, Galileo's discussion of his postulate of equal speeds (pp. 41–42, below).

[33] The principles which are made explicit include the principle of the balance and the law of the lever; nonetheless Galileo's language (in the mouth of Salviati, of course) implies that the science of resistance is founded on its own first principles. In fact, Sagredo asks for (and gets) a proof of the law of the lever so as to have "entire and complete instruction" (Drake, 1974a, p. 110). Galileo thus seems implicitly to deny that this is a subordinate science. In fact, I find him nowhere using language that suggest he thinks of his 'new sciences' as in any way imperfect, subordinate or mixed, in the sense

in which he remarks on such sciences in his early logical notes (see Wallace, in this volume; also Machamer, this volume).

[34] Clavelin sees Galileo's hesitation over the underlying cause of cohesion in the discovery of puzzles and contradictions in the concept of the infinite (1974, p. 446). I question, however, whether this is, in fact, the reason Galileo draws back from firmly asserting what he has conjectured about the nature of matter. His remark that reason must step in where the senses fail (see note 31 above) is made precisely in connection with this problem. Thus, it cannot be the limits of rationality alone that stop him; rather, it seems to be primarily the lack of the visual model which, in every instance, seems to be required for the establishment of fundamental principles.

[35] Drake (1974a, pp. 159–161); see also Drake (1970) for general discussion of Galileo's argument and review of much of the literature on this point. For the past few years, Drake has been urging that the rule $v:s$ was not mistaken (see, especially, Drake, 1974b). His argument is ingenious and helps exculpate Galileo from error. Unfortunately, however, it is vitiated by Galileo's own admission of guilt (Drake, 1974a, p. 160; *Opere* VIII, p. 203).

[36] In the fourth day of the *Two New Sciences*, Galileo proves that the maximum range of a projectile is obtained by firing at an elevation of 45° (Drake, 1974a, p. 245). But this result is not remarked upon as a confirmation of the principles from which it derived. Rather, it provides an occasion for celebrating the superiority of mathematical demonstrations over simple facts learned from experience. "The knowledge of one single effect acquired through its causes opens the mind to the understanding and certainty of other effects without need of recourse to experiments" (*ibid.*). The power of mathematics lies in the possibility of deducing new truths, that is, in the technique of 'demonstrative discovery' which Galileo emphasized in the *Discourse on Bodies in Water* (see Section II above and Wisan, 1974, pp. 120–125, 257–258).

[37] Most of the propositions on accelerated motion followed from the times-squared theorem (Theorem III) derived directly from the postulate. The propositions on projectile motion required the times-squared theorem, the definition of uniformly accelerated motion and the double-distance rule derived from that definition. Now, Galileo thought (for a time, at least) that he had an argument for the double-distance rule which would be independent of the assumption $v:t$ if $t^2:s$ were considered established by experiment alone, and in the book on projectile motion he actually derives $v:t$ from $t^2:s$ and the double-distance rule (Wisan, 1974, pp. 232–236, 227–228).

[38] *Opere* XVII, pp. 90–91; translated in Wallace (1976, p. 78). See similar remarks in *Opere* XVIII, pp. 11–12; translated in Wallace (*ibid.*, p. 88).

A quite different interpretation of this same material is suggested by Wallace, who seizes upon Galileo's use of the expression *ex suppositione* to argue that his actual method was an Archimedean modification of a medieval method of reasoning *ex suppositione* which was given its 'classical expression' by Aquinas (1976, p. 76; also see paper in this volume). Wallace's reconstruction, however, requires that Galileo's 'supposition' be an experimental result such as the times-squared theorem and his reasoning *ex suppositione* a derivation of the definition of uniformly accelerated motion (that is, $v:t$) from such results (1976, p. 92). But in no case cited by Wallace, or elsewhere so far as I know, does Galileo make such a derivation or use the expression in this way. Certainly, it is not so used in the third day of the *Two New Sciences* (see p. 40 above). Also, at the beginning of the fourth day, Galileo's explicit suppositions are the times-squared theorem, the inertial principle and the principle of independence of motions. All of these are then used to derive the parabolic path of projectiles, and there is no discussion of any experimental confirmation of principles. In the correspondence of 1637 and 1639, Galileo argues "*ex suppositione*, imagining for myself a motion. . .[which increases] its velocity with the same ratio as the time. . . ." (Wallace, 1976, p. 87), and again, "I assume nothing but the definition of the motion of which I treat. . . ." (*ibid*., p. 88). Thus, Galileo's use of the expression *ex suppositione* must be radically reconstructed to get the conclusions Wallace draws. Also, of course, Galileo did not have the methods of calculus available to him for reversing his mathematical derivation which proceeds from $v:t$ to $t^2:s$ as suggested by Wallace (*ibid*., p. 96, n. 7), and despite fresh evidence of various experimental results it remains unlikely that Galileo ever thought he had a demonstration of $v:t$ which was soundly based on such results (see Section I of this paper).

[39] C. Truesdell remarks on the general tendency of historians to neglect the development of the rational sciences, noting that "It is only in its application of mathematics to natural science. . . that the technology of the western culture differs in principle from others" (1968, p. 2).

[40] Throughout the age of reason, as Truesdell points out, rational mechanics was primarily the business of mathematicians, not of physicists; and it was generally felt that "rational mechanics, like geometry, must be based upon axioms which are obviously true," and that "further truth in mechanics follows by mathematical proof" (*ibid*., pp. 93, 94).

[41] Clavelin gives a somewhat similar explanation of how Galileo "came to be the first to take the great leap forward into mathematical physics," observing that "the crucial difference between him and his precursors" was that his fundamental law was a "mere stepping stone" from which he "proceeded to the systematic study of [the motion of heavy bodies] in the most diverse situations. . . ." (1974, p. 307). Clavelin also sees in Galileo an important substitution of "the principle of a genetic explanation" for "the descriptive and classificatory endeavors of traditional science" (*ibid*., p. 294). Clavelin is one of those very rare students of Galileo who stresses equally the empirical and the rational, the technical and the philosophical. His study is surely the most comprehensive and insightful that has yet been done on Galileo, at least since Koyré.

WORKS CITED

Cassirer, E.: 1922, *Das Erkenntnisproblem in der Philosophie und Wissenschaft der neueren Zeit*, 1. 2nd ed., Verlag Bruno Cassirer, Berlin (1st ed. 1906).

Clavelin, M.: 1974, *The Natural Philosophy of Galileo: Essay on the Origins and Formation of Classical Mechanics* (transl. by A. J. Pomerans), Massachusetts Institute of Technology Press, Cambridge (1st French ed. 1968).

Commandino, F.: (transl. and comm.), 1588, *Pappi Alexandrini Mathematicae Collectiones*. . . Pesaro.

Costabel, P.: 1975, 'Mathematics and Galileo's Inclined Plane Experiments', in *Reason, Experiment, and Mysticism in the Scientific Revolution* (ed. by M. L. Righini Bonelli and W. R. Shea), Science History Publications, New York.

Crescini, A.: 1965, *Le origini del metodo analitico: Il Cinquecento*. Del Bianco, Udine.

Crew, H. and de Salvio, A. (transl.), 1914, *Galileo Galilei: Dialogues Concerning Two New Sciences*, Dover, New York, unabr., unalt. repub. of 1st ed.

Crombie, A. C.: 1956, 'Galileo: A Philosophical Symbol', *Actes VIIIe Congrès International d'Histoire des Sciences* **3**, 1089–1095.

Crombie, A.C.: 1975, 'Sources of Galileo's Early Natural Philosophy', in *Reason, Experiment, and Mysticism in the Scientific Revolution* (ed. by M. L. Righini Bonelli and W. R. Shea), Science History Publications, New York.

Descartes, R.: 1961, *Rules for the Direction of the Mind* (transl. by L. J. Lafleur), Bobbs-Merrill, Indianapolis.

Drabkin, I. E. and Drake, S. (transl.), 1960, *Galileo Galilei: On Motion and On Mechanics*, University of Wisconsin Press, Madison.

Drake, S. (transl.), 1967, *Galileo Galilei: Dialogue Concerning the Two Chief World Systems – Ptolemaic and Copernican*, 2d ed., University of California Press, Berkeley (1st ed. 1953).

Drake, S.: 1970, 'Galileo Gleanings XIX: Uniform Acceleration, Space, and Time', *British Journal for the History of Science* **5**, 21–43.

Drake, S.: 1973, 'Galileo's Experimental Confirmation of Horizontal Inertia: Unpublished Manuscripts', *Isis* **64**, 291–305.

Drake, S. (transl.), 1974a, *Galileo: Two New Sciences Including Centers of Gravity and Force of Percussion*, University of Wisconsin Press, Madison.

Drake, S.: 1974b, 'Galileo's Work on Free Fall in 1604', *Physis* **16**, 309–322.

Drake, S.: 1975a, 'Galileo's New Science of Motion', in *Reason, Experiment, and Mysticism in the Scientific Revolution* (ed. by M. L. Righini Bonelli and W. R. Shea), Science History Publications, New York.

Drake, S.: 1975b, 'The Role of Music in Galileo's Experiments', *Scientific American* **232**, 98–104 (June).

Drake, S.: 1976, 'The Evolution of *De motu*', *Isis* **67**, 398–419.

Drake, S. and Drabkin, I. E. (eds. and transl.), 1969, *Mechanics in Sixteenth Century Italy*, University of Wisconsin Press, Madison.

Drake, S. and O'Malley, E. D. (eds. and transl.), 1960, *The Controversy on the Comets of 1618*, University of Pennsylvania Press, Philadelphia.

Drake, S. and MacLachlan, J.: 1975, 'Galileo's Discovery of the Parabolic Trajectory', *Scientific American* **232**, 102–110 (March).

Edwards, W. F.: 1960, 'The Logic of Iacopo Zabarella', Ph.D. dissertation, Columbia University.

Euclid: 1570, *The Elements of Geometrie of the Most Ancient Philosopher Euclide of Megara* [transl. and annotated by H. Billingsley with preface (and annotations) by John Dee], John Daye, London.

Fredette, R.: 1969, 'Les *De motu* "plus anciens" de Galileo Galilei: prolégomènes', Ph.D. dissertation, Université de Montréal.

Fredette, R.: 1972, 'Galileo's *De Motu Antiquiora*', *Physis* **14**, 321–48.

Galilei, G.: 1890–1909, *Le Opere di Galileo Galilei*. . . 20 vols. (ed. by Antonio Favaro), Tipografia di G. Barbèra, Florence.

Galluzzi, P.: 1973, 'Il "Platonismo" del tardo cinquecento e la filosofia di Galileo', in *Ricerche sulla cultura dell'Italia moderna* (ed. by P. Zambelli), Editori Laterza, Rome.

Giacobbe, G. C.: 1972a, 'Il *Commentarium de Certitudine Mathematicarum Disciplinarum* di Alessandro Piccolomini', *Physis* **14**, 162–93.

Giacobbe, G. C.: 1972b, 'Francesco Barozzi e la *Quaestio de Certitudine Mathematicarum*', *Physis* **14**, 357–74.

Gilbert, N. W.: 1960, *Renaissance Concepts of Method*, Columbia Univ. Press, New York.

Gilbert, N. W.: 1963, 'Galileo and the School of Padua', *Journal of the History of Philosophy* **1**, 223–31.

Heath, T. L. (transl.), 1956, *The Thirteen Books of Euclid's Elements...* Vols. I–III, reprint of 2d ed. 1926, Dover Publications, New York.

Hintikka, J. and Remes, U.: 1974, *The Method of Analysis: Its Geometrical Origin and its General Significance*, in Boston Studies in the Philosophy of Science, Vol. XXV (ed. by R. S. Cohen and M. W. Wartofsky), Reidel, Dordrecht.

Höffding, H.: 1955, *A History of Modern Philosophy* (transl. by B. E. Meyer), Dover, New York unabr. repub. of orig. Eng. transl. 1900.

Jardine, N.: 1976, 'Galileo's Road to Truth and the Demonstrative Regress', *Studies in History and Philosophy of Science* **7**, 277–318.

Klein, J.: 1968, *Greek Mathematical Thought and the Origin of Algebra* (transl. by E. Brann), Massachusetts Institute of Technology Press (1st German text 1934–1936).

Koertge, N.: 1977, 'Galileo and the Accidents', *Journal of the History of Ideas* **38**, 389–408.

Koyré, A.: 1939, *Étudès Galileennes*. I. 'A l'aube de la science classique.' II. 'La loi de la chute des corps: Descartes et Galilee.' III. 'Galilée et la loi d'inertie', *Histoire de la Pensée*, Nos. 852–54, Hermann, Paris.

McKeon, R.: 1965, 'Philosophy and the Development of Scientific Methods', *Journal of the History of Ideas* **27**, 90–110.

Moody, E. A.: 1951, 'Galileo and Avempace: The Dynamics of the Leaning Tower Experiment', *Journal of the History of Ideas* **12**, 163–93; 375–422.

Naylor, R. H.: 1974a, 'Galileo and the Problem of Free Fall', *The British Journal for the History of Science* **7**, 105–134.

Naylor, R. H.: 1974n, 'Galileo's Simple Pendulum', *Physis* **16**, 23–46.

Naylor, R. H.: 1974c, 'The Evolution of an Experiment: Guidobaldo del Monte and Galileo's Discorsi Demonstration of the Parabolic Trajectory', *Physis* **16**, 323–346.

Naylor, R.H.: 1976a, 'Galileo: The Search for the Parabolic Trajectory', *Annals of Science* **33**, 153–172.

Naylor, R. H.: 1976b, 'Galileo: Real Experiment and Didactic Demonstration', *Isis* **67**, 398–419.

Panofsky, I.: 1968, *Idea: A Concept in Art History* (transl. by J. J. S. Peake from rev. German ed. 1960), Harper and Row, New York. (1st German ed. 1924.)

Popkin, R.: 1960, *The History of Scepticism from Erasmus to Descartes*, Van Gorcum, Assen.

Poppi, A.: 1972, *La dottrina della scienza in Giacomo Zabarella*, Editrice Antenore, Padova.

Proclus: 1970, *A Commentary on the First Book of Euclid's Elements* (transl. with Int. and Notes by G. R. Morrow), Princeton University Press, Princeton.

Randall, J. H., Jr.: 1961, *The School of Padua and the Emergence of Modern Science*, Editore Antenore, Padua. (Expanded version of 'The Development of Scientific Method in the School of Padua', *Journal of the History of Ideas* **1**, 177–206.)

Rose, P. L.: 1976, *The Italian Renaissance of Mathematics*, Librairie Droz, Geneva.

Salusbury, T. (transl.), 1960, Galileo Galilei: *Discourse on Bodies in Water* (with Int. and

Notes by S. Drake), University of Illinois Press, Urbana.
Schmitt, C. B.: 1969, 'Experience and Experiment: A Comparison of Zabarella's View with Galileo's in *De Motu*', *Studies in the Renaissance* **16**, 80–138.
Schmitt, C. B.: 1972, 'The Faculty of Arts at Pisa at the Time of Galileo', *Physis* **14**, 243–72.
Schmitt, C. B.: 1973, 'Towards a Reassessment of Renaissance Aristotelianism', *History of Science* **11**, 159–193.
Settle, T. B.: 1961, 'An Experiment in the History of Science', *Science* **133**, 19–23.
Settle, T. B.: 1966, 'Galilean Science: Essays in the Mechanics and Dynamics of the *Discorsi*,' Ph. D. dissertation, Cornell University.
Settle, T. B.: 1967, 'Galileo's Use of Experiment as a Tool of Investigation', in *Galileo Man of Science* (ed. by Ernan McMullin), Basic Books, New York.
Settle, T. B.: 1968, 'Ostilio Ricci, a Bridge between Alberti and Galileo', *Actes XIIe Congrès International d'Histoire des Sciences* 3b, 121–126.
Shea, W. R.: 1972, *Galileo's Intellectual Revolution: Middle Period, 1610–1632*, Science History Publications, New York.
Sosio, L. (ed.): 1965, *Galileo Galilei: Il Saggiatore*, Feltrinelli, Milano.
Sosio, L. (ed.): 1970, *Galileo Galilei: Dialogo sopra i due massimi sistemi del mondo, tolemaico e copernicano*, Einaudi, Turin.
Sosio, L.: 1971, 'I "Pensieri" di Paolo Sarpi sul moto', *Studi Veneziani* **13**, 315–92.
Trusedell, C.: 1968, *Essays in the History of Mechanics*, Springer, New York.
Wallace, W. A.: 1974, 'Galileo and the Thomists', *St. Thomas Aquinas Commemorative Studies* 2, 293–330, Pontifical Institute of Medieval Studies, Toronto.
Wallace, W. A.: 1976, 'Galileo and Reasoning *Ex Suppositione*: The Methodology of the *Two New Sciences*', in Boston Studies in the Philosophy of Science, Vol. XXXII. (Proceedings of the Philosophy of Science Association 1974) (ed. by R. S. Cohen *et al.*), Reidel, Dordrecht and Boston, p. 79.
Wisan, W. L.: 1974, 'The New Science of Motion: A Study of Galileo's *De motu locali*', *Archive for History of Exact Sciences* **13**, 103–306.

ROBERT E. BUTTS

SOME TACTICS IN GALILEO'S PROPAGANDA FOR THE MATHEMATIZATION OF SCIENTIFIC EXPERIENCE

It has frequently been claimed that Galileo is the father of modern science. Historians of science who thus enshrine him claim for him nor only important scientific discoveries, but also the discovery of the telescope, the introduction of the first genuine scientific method, defined mainly by reliance on experimentation, and the destruction of the prevailing Aristotelian metaphysics of his day. Other writers proclaim him for his restoration of Platonism (Koyré, 1939, 1943), or condemn him for introducing the basic elements of a subjectivist view of man destined to lead to no good (Burtt, 1932). In the past few years, we have begun to get new motivation to think about Galileo by new assessments of his work, some balanced (Shea, 1972), others highly skewed and controversial (Feyerabend, 1970). Perhaps it is somewhere between the clear historico-philosophical analysis by Shea of Galileo's achievements, and Feyerabend's contention that Galileo was one of the greatest propagandists of ideas in the history of science that the truth about Galileo lies. Certainly enough is now know about this great man of science to realize that it would be a mistake to pin him up in a museum case as just one more 'father.'

I will, therefore, begin by conceding what seems obvious: Galileo did indeed make a substantial contribution to positive science, even though he frequently did not understand his own contribution, as in the case of the acceleration-rate law. It also seems to me true that although Galileo did not invent experimentation, he did most importantly modify the epistemological point of doing experiments.[1] Finally, unlike many scholars now talking about Galileo, I for one find plenty of evidence that – even though in some respects he was an Aristotelian – his system is a massive attack upon a prevailing stereotype of Aristotle quite characteristic of the age in which he lived. Briefly reviewed, 'this' Aristotle held that mathematical and physical descriptions are distinct and that only physical explanations count; the universe is an organism striving to be like the God who created it; in the order of being man stands closest to God of all created beings; this metaphysical centrality of man finds

R. E. Butts and J. C. Pitt (eds.), New Perspectives on Galileo, 59–85.

its physical complement in the fact that man's Earth is the literal centre of the physical universe – a self-contained package which when fully worked out gives a justification for a certain way of construing the Aristotelian texts in a wholly coherent way.

Even given all of this I also believe that we must concede to Feyerabend the implied point that Galileo was not much of a philosopher *in a certain sense*. Galileo's programme to bring mathematics back to the throne, for example, cannot be regarded as a philosophical achievement in the sense that Galileo developed a rational reconstruction of even his own science in such a way as to give a metaphysical and epistemological justification of his 'new' way of doing science. Galileo was not Descartes; nor was he Leibniz. It is the entire array of his writings, his political activities, his involvement in religious and theological quarrels, in short, it is the message of his entire life that stands against the Aristotelian stereotype. Koyré and Shea are in part mistaken. Galileo was not a Platonist without qualification, and what qualifies his view is here of the utmost importance. I will try to tell some parts of this story as we progress. Maybe at this point it will suffice to suggest that Galileo's metaphysics is underdeveloped and hence that both philosophical argument and systematic development of themes come forth from his works as polemical challenges, but not as integrated philosophy as the major figures in the 17th century understood it. We should not look for an articulated integration of philosophy and science of the kind that we find in Descartes and Leibniz in the works of Galileo. His is a different style: suggestion, rhetoric, use of the argumentum ad hominem, appeals to Aristotle that are twisted into defenses of Copernicanism – all of this (and more) strongly urges Feyerabend's interpretation upon us. At the very least it might be of interest to have a look at this new perspective on Galileo. If one is prepared to make the concessions that I have made the ground is prepared.

In this place I can only look at one aspect of Galileo's programme. I have deliberately chosen the central part of that programme: the attempted reduction of scientific experience to experience that can be expressed in mathematical terms. This part of the programme appears to be simple and to be established with some ease. It also appears to exhibit Galileo as the17th century metaphysician par excellence. We will see, however, that the arguments are not very strong, and that the distinctions required are not at all convincingly made. Feyerabend seems to be right – the Galileo whose posture we are about to study is not so much a philosopher as he is a propagandist, not so much a scientist as he is a metaphysical politician.[2]

Galileo's claim for the supremacy of mathematical knowledge is stated with great conviction, a conviction that is somewhat startling for, among other implications, its apparently deviant theological consequences.

> . . .human understanding can be taken in two modes, the *intensive* or the *extensive*. *Extensively*, that is with regard to the multitude of intelligibles, which are infinite, the human understanding is as nothing even if it understands a thousand propositions; for a thousand in relation to infinity is zero. But taking man's understanding *intensively*, in so far as this term denotes understanding some propositions perfectly, I say that the human intellect does understand some of them perfectly, and thus in these it has as much absolute certainty as Nature itself has. Of such are the mathematical sciences alone; that is, geometry and arithmetic, in which the Divine intellect indeed knows infinitely more propositions, since it knows all. But with regard to those few which the human intellect does understand, I believe that its knowledge equals the Divine in objective certainty, for here it succeeds in understanding necessity, beyond which there can be no greater sureness. (Galileo Galilei, 1632, p. 103.)

Galileo's confidence in the power of the human intellect involves epistemologically curious implications. Against the Aristotelians, he wished to hold that mathematics is not just formal description (implying that many mathematical descriptions of a given phenomenon are possible, but that only one true physical description is possible), but that physical nature is somehow ultimately mathematical. At the same time, Galileo, also against the Aristotelians, introduced the first major campaign in favor of physical experimentation. The puzzle is that one apparently does experiments in order to find out how nature actually works; experimentation is a kind of manipulated observation. But what is it that guarantees that an experiment will reveal *only* mathematical realities? Ought one not to say: "Test it and see, that's the only way science can learn what it's all about"? Following a certain way or reading Plato – and certainly this way has from time to time been popular – one might argue that *reality* is mathematical, but that reality is not that which is revealed as the result of experimentation. Whatever one's ontological proclivities are, one can see that this reading is at least an important candidate among the various ontological choices. We do not observe triangles and circles and threes, but that's all right, what we observe is not, for various reasons that need argument, what is scientifically real. Galileo was not, I think, the kind of Platonist just alluded to. For he wanted what we observe as the result of experiments to count as real; at the same time he wanted that reality to be mathematical.

I have been discussing Galileo's view as expressed in the *Dialogue*, published in 1632. It must be noted, however, that this view is not the same as that expressed in earlier works, for example, *History and Demonstrations concerning Sunspots and their Phenomena* (commonly referred to as 'Letters on Sunspots') published in Rome in 1613. In the 'Letters' Galileo appears to adopt a relatively straightforward form of exactly the kind of Aristotelianism that he would later combat. In one section he accuses the mysterious Apelles of continuing to

> adhere to eccentrics, deferents, equants, epicycles, and the like as if they were real, actual, and distinct things. These, however, are merely assumed by mathematical astronomers in order to facilitate their calculations. They are not retained by philosophical astronomers who, going beyond the demand that they somehow save the appearances, seek to investigate the true constitution of the universe – the most important and most admirable problem that there is. For such a constitution exists; it is unique, true, real, and could not possibly be otherwise. . . (Galileo Galilei, 1613, pp. 96–97.)

In the second Letter Galileo reinforces this apparent Aristotelianism, writing

> I should even think that in making the celestial material alterable, I contradict the doctrine of Aristotle much less than do those people who still want to keep the sky inalterable; for I am sure that he never took its inalterability to be as certain as the fact that all human reasoning must be placed second to direct experience. Hence they will philosophize better who give assent to propositions that depend upon manifest observations, than they who persist in opinions repugnant to the senses and supported only by probable reasons. (*Ibid.*, p. 118.)

Of course the 'Letters' are polemical pieces, although they contain much that is of positive scientific interest. Nevertheless, the philosophical position stated by Galileo in the 'Letters' is by no means compatible with his philosophy in the *Dialogue* and in other later writings. At this point Feyerabend would applaud Galileo for his boldness in holding conflicting opinions, and would recommend that other scientists follow his example. But Galileo held these opinions at different times, so that perhaps the explanation lies in the fact that he simply changed his mind. Attractive as this alternative is, I do not think it adequately accounts for the apparently different views that Galileo stated.

Notice that in the first quote Galileo criticizes those who think of geometrical astronomy as a science whose aim is merely to 'save the appearances.'

This was, I think, never Galileo's view of astronomy, even in early stages of his career. Astronomy, like any proper science, reveals the real world, as he goes on to point out in the quoted passage. All that is lacking from this earlier statement is Galileo's expression of his belief that the fixed constitution of the real world is mathematical in character. The difference is simply between taking mathematics to be a kind of aesthetic pastime and taking mathematical descriptions to be real – in opposition to Apelles and his ancestors. The second quotation is also basically neutral on the question of what kind of observables are to be the confirmatory materials of science. Galileo can agree with Aristotle that observation is more to be trusted than abstract reasoning. At the same time, he can disagree with the Aristotelians of his time that observation must be uncontrolled, 'natural' observation. As we will see, Galileo thought that the observational results of a good experiment must always be assertable in quantitative terms. These passages from Galileo's earlier polemical writings do not then seem to conflict in any logically compelling way with his later, more fully stated views.

Even granting this reading of the early texts, Galileo's philosophical task must be viewed as being enormous. Believing as he did that genuine knowledge of the physical world is mathematical in character, and believing as he did in the efficacy of experimental science, he was faced with having to show what appears to be counter-intuitive, namely, that experimental observables are numbers or geometrical shapes, that the very experiences that experiments bring about are mathematical experiences. To help in understanding Galileo's problem, we might consider that a couple of centuries later Kant addressed himself to the same problem, but in epistemological rather than ontological terms. For Kant, to talk about the presuppositions of all possible experience was in large part to show how scientific experiences must be mathematical. The important difference is one of philosophical emphasis. Kant's approach to this problem was almost entirely epistemological; his major concern was with analysing what is required in order for us to have scientific knowledge. For purposes of this analysis Kant's ontology was very thin. Not so for Galileo. He appears to have thought that the question of justifying mathematical knowledge required introducing a quite substantial ontology. On this way of moving we are sure that a certain way of knowing is justified because we have direct access to the reality of the kind of thing known. Independent of our scientific knowing we know that this form of knowing is the only correct one. The question is: did Galileo accomplish the job required to show that we have this independent access? The question, moreover, is by no means idle. For the knowledge that scientific knowledge is supreme is not itself scientific knowledge.

Galileo understood the philosophical problem as stated above, but unfortunately he did not set out an extensive philosophical defense of his ontology. Elements of this ontology are scattered throughout his works. It is unlikely, however, that an unforced coherent reconstruction of his position can be given. When talking about Galileo it is at junctures like this that one feels the force of Feyerabend's suggestion that Galileo was a superior propagandist for his ideas. Galileo does after all introduce his form of the distinction between primary and secondary qualities, and he does 'argue' in the *Dialogue* that what is true is geometry is true of the physical world. One might even regard these two portions of his thought as essential for understanding what he thought could be expected from experiments in physics. As will be seen, however, Galileo does not put together an unambiguous or powerful case for this conclusion. It seems preferable, therefore, to view the secondary-primary qualities distinction and the argument in the *Dialogue* as tactical moves in a continuing polemic against those who would reject the new philosophy of science. Let us proceed to look at Galileo's distinction and argument in this new way.

Galileo's distinction between the two kinds of qualities has often been quoted, and often discussed (usually, one might add, in a somewhat distorted way). Those familiar with Galileo's treatment of the distinction will I hope forgive me for quoting it once again for convenience of reference.

> Now I say that whenever I conceive any material or corporeal substance, I immediately feel the need to think of it as bounded, and as having this or that shape; as being large or small in relation to other things, and in some specific place at any given time; as being in motion or at rest; as touching or not touching some other body; and as being one in number, or few, or many. From these conditions I cannot separate such a substance by any stretch of my imagination. But that it must be white or red, bitter or sweet, noisy or silent, and of sweet or foul odor, my mind does not feel compelled to bring in as necessary accompaniments. Without the senses as our guides, reason or imagination unaided would probably never arrive at qualities like these. Hence I think that tastes, odors, colors, and so on are no more than mere names so far as the object in which we place them is concerned, and that they reside only in the consciousness. Hence if the living creature were removed, all these qualities would be wiped away and annihilated. But since we have imposed upon them special names, distinct from those of the other and

real qualities mentioned previously, we wish to believe that they really exist as actually different from those. (Galileo Galilei, 1623, p. 274.)

To excite in us tastes, odors, and sounds I believe that nothing is required in external bodies except shapes, numbers, and slow or rapid movements. I think that if ears, tongues, and noses were removed, shapes and numbers and motions would remain, but not odors or tastes or sounds. The latter, I believe, are nothing more than names when separated from living beings, just as tickling and titillation are nothing but names in the absence of such things as noses and armpits. (*Ibid.*, pp. 276–77.)[3]

Much nonsense has been written about what Galileo has to say in these two passages and surrounding text. In Drake's note to his translation of the first passage, he endeavors to defend Galileo against the suggestion that his distinction anticipates the empiricism of Locke and others. Drake's line of approach seems irrelevant to an understanding of the importance of the distinction, especially since the distinction – or something very much like it – appears in the works of others regarded as *rationalists* in the more-or-less standard philosophical tradition.[4] E. A. Burtt, in his important and well-known book *The Metaphysical Foundations of Modern Science*, has misled almost two generations of readers by fastening attention upon the apparent *subjectivity* of the sensory qualities. (Notice that Galileo does not use this nasty word; apparently 'reside only in the consciousness' is identical with 'subjective' for Burtt and others.) The term 'subjective,' of course has unfortunate connotations, some of which lead Burtt to the theologico-metaphysical conclusion that Galileo's distinction reads man and his experience out of the world of the real. (Burtt, 1932, pp. 89–90.) Galileo's distinction may indeed have had for some thinkers the effect of introducing a dualistic ontology that relegated man to a minor status in the universe. For others, stressing the apparent nominalism of Galileo's distinction, the reality of man's *experiences* might be called specifically into question. But Galileo did not *say* any of this, nor was it his clear intention ontologically to dethrone man. After all, he was trying to show how it is that in the case of mathematics man's understanding equals that of God!

So perhaps one can be excused for taking a new and closer look at the hoary corrupter of the then-new science. The distinction seems to me to involve the following claims:

(1) The properties of *matter* are essentially mathematical, i.e., measurable.

(2) Sensory qualities are not *in* matter, but exist only in consciousness.

(3) Sensory qualities are no more than names '*so far as the object in which we place them is concerned.*'

(4) To exist, sensory qualities require a sense organ *and* an external material stimulus.

(5) For the purpose of doing *publically available* science, private and idiosyncratic sensory qualities are irrelevant *because* they are not mathematizable.[5]

Notice that Galileo's position as expressed in the five points above is literally ontological. Proposition (1) characterizes matter in a certain definite way, but it does not entail that *only* matter is real. (2) is also clearly ontological; it locates sensory qualities in the consciousness of those who sense things. And (3) tells us, for example, that when we say 'this fire is hot,' we are to be understood as saying 'when I touch this fire, I have a sensation of heat.' (2) and (3) do not commit Galileo to a nominalism with respect to the secondary qualities, but only to a certain semantical programme for understanding the meanings of certain phrases in ordinary language. It is obvious that if the sensory qualities exist only in consciousness then sensory words like 'sweet,' 'red,' 'bitter,' and the like are empty of reference ('mere names') when applied to physical objects. Of course Galileo misses a good deal in laying down this ontology: there may still remain perfectly good senses in which I can claim 'this x is red' and take the sentence to be publically verifiable. But niceties of epistemological analysis were far from Galileo's mind in championing this position.

Proposition (4) introduces the notorious 'propensity' theory according to which we are to regard material objects as having the potentiality to create sensations in us. But again, (4) does not deny the reality of sensory qualities, nor does it necessarily relegate them to a lower ontological status. Galileo's position leaves it quite open that the mind might come upon other ideas (again the obvious comparison is with the thought of Descartes) that are not sensory in origin, and evidently do not need the presence of physical objects as conglomerates of mathematical objects as their causes. Proposition (5) must be seen as the key to Galileo's purposes in introducing the distinction. He notes that if I move a feather over a statue and in the same manner over a closely similar naked living body only the living body will – at some points – feel the sensation of tickling. Tickling appears as a scientifically recalcitrant fact about certain kinds of bodies. But in the two cases what is publically observable is the same motions of the feather, and the components of that motion can be expressed geometrically. We are now, I think, into the heart of the matter. (1), (2), and (3) guarantee that in the two cases at hand we have two distinct kinds of object: the geometrically determinable motion of the feather over the surfaces of the two bodies, and the private sensation of tickling in the living body. Both are real; I see no reason for thinking that it was Galileo's intention to suggest that the motions are 'more real' than the sensations.

Unhappily for his case, Galileo's argument from analogy is not strong enough to support the conclusion that he needs. The similarity between sensations like tickling and perceptions of sound and color is not so striking as he thought it was. Part of the argument appeals to ordinary usage: we would not *say* that the tickling is in the feather that tickles. 'I am tickled' is the acceptable locution. On the other hand, it is part of ordinary language to say, and mean it, that 'this water is hot,' and that 'this book is red.' We cannot argue from the tickling case to the heat or color or sound cases, certainly not by appeal to ordinary usage. In any case, Galileo needs to establish (3) for *all* sensations; what he gets from the quite inadequate argument from analogy is that *maybe* there are sensations that are different in kind, those that ordinary language locates in the perceiver, and those that common usage (wrongly) attributes to the external material object. In addition, Galileo misses the crucial point that sensations are measurable on scales that mark degrees of intensity, and that we can, given such measure scales, determine, say, the heat of an object by taking its temperature, a process that requires no sensations at all. When a body of water measures 212° Fahrenheit it is certainly quite inadequate to say that *the water itself is in no sense hot*, but that it would be if someone plunged his hand into it. Motion of molecules and an increase of volume in a certain body of water is crucially required in order to produce the sensation of heat in someone who performs this experiment, which proves Galileo's (4). It by no means proves (2) and (3), just those propositions required to have the distinction between primary and secondary qualities.

Thus Galileo can be right that motions are the *cause* of heat, and still be wrong that the heat in no sense exists *in* the object, e.g., the boiling water. Certainly the thermometer measures *something*, and it is not a something that exists merely as a potentiality to produce a sensation of heat in a perceiver. What the thermometer measures it is the very difficult business of physics to find out, and whatever that something is is intensity of heat (or intensity of some other quality). The argument can be extended beyond the very simple cases that Galileo considers. Even felt sensations are susceptible to kinds of measurements – the concept of pain thresholds may be taken as a case in point. However, it would be unfair to expect him to talk to the point of such extensions given that the Aristotelian empiricism that he was combatting required important revisions even to get experimentation of any kind started. His unsupported philosophical distinction had at least the virtue of opening the way to seeing that nature can be expected to give answers if manipulated in controlled ways. To introduce a new alternative position is not necessarily to be successful in establishing it.

Leaving related and residual epistemological and analytical problems aside, I return to what I take to be the more lasting contribution implicit in Galileo's form of the distinction between the two types of qualities, a result that forms a basic part of proposition (5). Galileo was doing physics at least in large part in a new style; as part of the philosophical programme required to justify his new methodology he needed a characterization of physical objects. The ontology implicit in his distinction can be seen to give way to a more basic methodological requirement. Physics is not about what we see by naked-eye observations, nor about any of our private sensations – it is about what is publically available with respect to bodies, i.e., it is about those mathematical properties that make bodies capable of being measured in certain ways, and 'observed' in experimental contexts. I think that Shea's conclusion following his analysis of Galileo's distinction is exactly right:

> . . .since the nature of bodies could never be fully understood from what was perceived of their behavior, uninterpreted (that is, non-geometrised) facts were only the raw materials of science. The senses were no longer reliable guides to knowledge, and the role of experiment was to confirm or refute what had been deduced from geometrical considerations. (Shea, 1972, p. 105.)

Seen as a methodological regulation of what we can expect from experiments, Galileo's distinction perhaps avoids many of the epistemological and (supposedly) theological problems attendant upon the distinction as Locke and others tried to use it. However, not all of the story has been told. It is one thing to insist that experimental observables must conform to a certain methodological principle; it is quite another to get all physicists to accept the new principle. One cannot generate an acceptable concept of scientific experience by methodological fiat alone, some argument is needed. Galileo thought that he had been successful in providing the required argument. I now turn to the details of this argument, as tricky a piece of metaphysical manoeuvring as one can find anywhere.

I think it must be said that my reading of Galileo's purpose in introducing the distinction between two kinds of qualities does not reduce it to a completely arbitrary methodological principle. What I sought for was a shift in emphasis. Galileo was not so much concerned with reducing the importance of human beings as with stressing the neglected prominence of certain features of the world – the fundamental nature of shape and motion, and hence the fundamental place of geometry as the language of nature. Almost no one, not even some of the most thoroughly indoctrinated Aristotelians denied that one can do geometry and that the world in some sense confirms a geometrical description. Galileo's treatment of the distinction between two kinds of qualities does not so much denegrate sensory qualities as restore physical objects to a status they had deserved but seldom attained. To deny that heat is in fire is incisively to deny that a certain substance has a certain accident. It is known that Galileo had little time for the doctrine of occult qualities; I am now suggesting that he had even less time for Aristotle's *grammar of physics*. The substance-accident way of structuring talk about physical objects leads one to distrust experimental manipulation of nature: leave the thing alone, its behaviour will tell you all you can know about it *au naturel*. It is not surprising that Aristotelians got the wrong answers; they were too satisfied with the results of naked-eye observations. But the naked-eye observations were not so naked as might at first appear. Aristotelian talk about such observations was already structured. We are watching the relatively permanent undergo changes, and its changes are already in some mysterious way contained in it.

This is not to say that there is not as much mystery in Galilean experimentation as there is in Aristotelian let-nature-take-her-course. Galileo's distinction between the two kinds of qualities had, however, exposed a vital nerve. His challenge consisted in jockeying the opposition into an intolerable set of alternatives. I give you a way of treating physical objects and their behaviour that is public and (in the best cases) relatively accurately measurable. Your alternatives are two: continue the fatiquing and exasperating appeal to authority, or show me how an intersubjectively verifiable physics can be built up on the basis of appeal to undisturbed experiences, given that such experiences have just that uncontrolled idiosyncracy that makes them unfit as candidates for functioning as the basis of scientific findings. There is, of course, a respect in which Galileo's positive scientific achievements vindicated his ontology and methodology; there is also a respect in which those achievements can never be taken as doing the philosophical job of vindication all alone.

So Galileo went to the heart of the matter. The problem had been posed as one between Platonists and Aristotelians, the former believing that nature is mathematical in character, the latter believing that mathematical descriptions are neither true nor false, but can only be judged on extraevidential criteria (e.g., aesthetic considerations), while a physics in the substance-accident grammar told us the truth about the observed physical world. To break the magical hold of this privileged grammar, Galileo had to attempt the improbable. He had to address himself directly to the question of the applicability of mathematics – in this case geometry – to the world as perceived by Aristotelians. His distinction between primary and secondary qualities had exposed the nerve; he now attempted to cut it fatally.

In the *Dialogue* Simplicio puts the appropriate Aristotelian point:

> . . .these mathematical subtleties do very well in the abstract, but they do not work out when applied to sensible and physical matters. For instance, mathematicians may prove well enough in theory that *sphaera tangit planum in puncto*. . . ; but when it comes to matter, things happen otherwise. What I mean about these angles of contact and ratios is that they all go by the board for material and sensible things. (Galileo Galilei, 1632, p. 203.)

The opposition has placed the appropriate hurdle; Galileo must jump high and far. Put without metaphor, Galileo is now faced with solving the problem of how it is that geometry *applies* to the world. The move is fair. It is Galileo who wants to hold that physical reality is geometrical. Simplicio acknowledges that he believes that a tangent would not in an actual physical situation touch the surface of the earth in only one point, but in many points along its surface. Salviati points out that this would mean that it is impossible for a projectile ever to leave the surface of the earth, since the angle of projection is completely closed because the tangent is united with the surface. Salviati's argument is not completely convincing. Simplicio had not suggested that the tangent would strike points along the total surface of the earth in a given direction. The best that Salviati could get from Simplicio's admission is the conclusion that the projectile would get off to a very bumpy start somewhat in the nature of a badly flown aircraft on takeoff![6]

Sophistries dispensed with, Salviati is now ready to get down to fundamental considerations. He teases Simplicio on the matter of the definition of a sphere, opening with:

Now to show you how great the error is of those who say, for example, that a sphere of bronze does not touch a steel plate in one point, let me ask you what you would think of anyone who might say – and stubbornly insist – that the sphere was not truly a sphere?" (Galileo Galilei, 1632, p. 204.)

Simplicio is caught off guard by this query, but after all it was he who insisted that what is true in geometry is not true of physical objects. So the question is fair, and the implication of the question is puzzling. Galileo is suggesting that in those cases where actual material objects do not obey the geometrical theorem, the material objects involved simply do not satisfy the geometrical definitions of (in this case) sphere and point. Simplicio might have replied that Salviati wants it both ways: the geometry must apply, otherwise the objects involved simply are not geometrical objects. There is, however, another obvious option: bronze spheres and steel plates might be taken as *falsifying* the theorems of geometry. Simplicio misses this move (the interesting thing is that Galileo allows him to miss this move), and is drawn into the matter of definition that he cannot win. He grants that what counts as a sphere is an object in which all straight lines drawn from its centre to its circumference are equal. *Given* this agreed-upon definition, Simplicio has just lost his first point. If the steel plate touches the bronze sphere along a portion of the surface of the latter, then the definition of a sphere rules out the bronze object as a sphere. The two solid objects produce a kind of physical perturbation along portions of their surfaces, and hence they cannot be treated as geometrical objects at all. *If they were* spheres and plane surfaces, then the material objects would obey the geometrical theorem. As we will see, this apparently innocent counterfactual will assume great importance in the position that Galileo will eventually evolve as a result of his developing argument.

Salviati, not content with a victory based upon matters of definition, immediately pursues another line. He gets Simplicio to agree that the straight line is the shortest line that can be drawn between two points. The discussion continues:

. . .And as to the main conclusion, you say that a material sphere does not touch a plane in a single point. Then what contact does it have?

Simplicio: It will be part of the surface of the sphere.

Salviati: And likewise the contact of one sphere with another equal one will still be a similar potion of its surface?

Simplicio: There is no reason that it should not be.

Salviati: Then also the two spheres will touch each other with the same two portions of their surfaces, since each of these being adapted to the same plane, they must be adapted to each other.

Now imagine two spheres touching whose centres are A and B, and let their centres be connected by the straight line AB passing through their contact. Let it pass through the point C, and take another point D in this contact, connecting the two straight lines

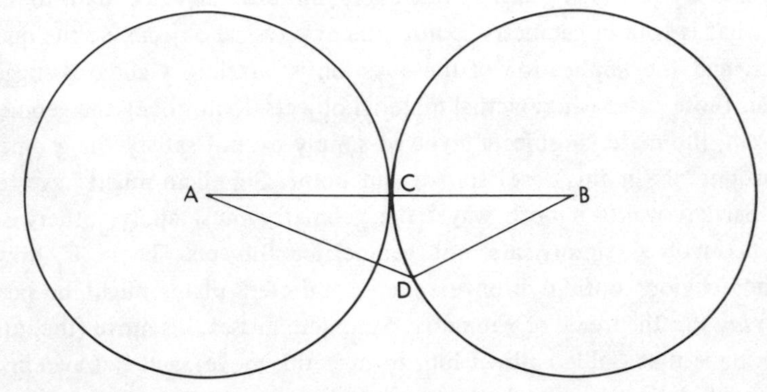

AD and DB so that they form the triangle ADB. Then the two sides AD and DB will be equal to the other single side ACB, each of them containing two radii, which are all equal by definition of the sphere. And thus the straight line AB drawn between the two centres A and B will not be the shortest of all, the two lines AD and DB being equal to it; which you will admit is absurd.

Simplicio: This proves it for abstract spheres, but not material ones. (*Ibid*., pp. 205–6.)

Thus Simplicio remains unconvinced, even by this new and ingenious argument. He points out, rightly, that material objects are subject 'to many accidents' and that the same is not true of immaterial (geometrical) objects. Matter is porous, for example, and thus no perfect spheres are to be found in nature. To all of this Salviati assents, following which he extracts the real point that his geometrical argument had actually masked:

. . .When you want to show me that a material sphere does not touch a material plane in one point, you make use of a sphere that is not a sphere and of a plane that is not a plane. By your own argument, spheres and planes are either not to be found in the world, or if found they are spoiled upon being used for this effect. It would therefore have been less bad for you to have *granted the conclusion conditionally; that is, for you to have said that if there were given a material sphere and plane which were perfect and remained so, they would touch one another in a single point, but then to have denied that such were to be had.* (*Ibid.*, pp. 206–7; emphasis added.)

Apparently Salviati's geometrical argument is window-dressing. Certainly the point just made can be regarded as quite independent of this argument. The new point emerges from a simple consideration of whether or not the original geometrical theorem holds for material objects. In any case, Salviati's new point needs careful consideration. It is one of the two most central theses that Galileo will argue for in his attempt to show that mathematics is the language of nature.

The conditional to which Salviati directs our attention is curious. Let us reformulate it as follows:

(C) For any x, y and t, if x is a perfect material sphere and y is a perfect material plane, and t is a definite interval of time, and x and y remain perfect through t, then x and y touch one another in a single point when y is struck as a tangent of x.

In appearance, (C) is a truth of geometry in applied form, that is, it is an empirical claim about the world. As such it ought to be falsifiable. But Galileo will not allow this possibility of falsification. Instead, he reads the conditional as a special counterfactual *that will always be true*, given that the antecedent can never be satisfied by actual solid objects. The subsequent discussion in the *Dialogue* makes Galileo's commitment to this analysis of the conditional clear.

Simplicio agrees that Aristotle's position is correctly formulated in this conditional manner, but then misses Salviati's point altogether by suggesting that material imperfections prevent things in the world from corresponding to abstract geometrical objects. Salviati jumps on this suggestion, holding that one who accepts (C) is in fact agreeing that material and geometrical objects 'exactly correspond.'

Salviati: Are you not saying that because of the imperfection of matter, a body which *ought to be* perfectly spherical and a plane which *ought to be* perfectly flat do not achieve concretely what one imagines of them in the abstract?

Simplicio: That is what I say.

Salviati: Then whenever you apply a material sphere to a material plane in the concrete, you apply a sphere which is not perfect to a plane which is not perfect, and you say that these do not touch each other in one point. But I tell you that *even in the abstract*, an immaterial sphere which is not a perfect sphere can touch an immaterial plane which is not perfectly flat in not one point, but over a part of its surface, so that what happens in the concrete up to this point happens the same way in the abstract. . . (*Ibid.*, p. 207; emphasis added.)

The insistence that material imperfections keep solid objects from being 'what they ought to be' lends greater force to the interpretation of (C) as a sacrosanct counterfactual. Salviati's clincher, of course, is his admission that what happens in the world is equivalent to what happens in geometry. This move completely overturns Simplicio's appeal to material imperfections as the reason why mathematics does not apply in the concrete. The question remains however: just what kind of victory has Galileo won?

It seems too easy to accept (C) as a true counterfactual; one continues to have the uneasy feeling that the problem – how does mathematics get to be applied to the world? – remains unsolved. Moreover, it was Galileo's intention to prove that mathematics is the privileged language in which to do science. The so-called material imperfections must therefore be dealt with. Otherwise it will appear that Galileo has secured a victory for mathematics in a *purely conventionalist manner*. Realizing this problem, Galileo has Salviati point out that in applying mathematics to the world, one "must deduct the material hindrances," and having done so, one will see that what is true in geometry is also true in the world.

The thesis of deducting the material hindrances is the second central proposition in Galileo's discussion of the applicability of mathematics to the world. Galileo does not argue for this thesis, and the missing argument raises important questions. One wonders about the apparent fiddling with the data that this thesis encourages, and one wonders about the *warrant* for such tampering. An additional question is how it is that Galileo thought that this unsubstantiated move proved that geometry is the supreme scientific language. These questions become all the more pressing when it is remembered that more recent philosophers of science have used something like Galileo's tactic to underwrite the conventionalism of geometry, a conclusion that I take to be the very opposite of what Galileo wanted to hold. The thesis of deducting the material hindrances (call it DMH) therefore needs a more careful look. I think it will emerge that the combination of DMH and (C) leaves Galileo in a *philosophically* hopeless position so far as justifying confidence in applied mathematics is concerned. However, as with the distinction between two kinds of qualities, Galileo may be seen as getting a good deal of *methodological* mileage out of combining DMH and (C).

Galileo draws an analogy. Consider a warehouse inventory clerk interested in knowing how much sugar he has in stock. For the purpose of determining this amount he neglects the contingent facts that some of his sugar is in bags, some in boxes, some in open containers ready to be sold retail. If what he wants, say, is a number in pounds, things like the sizes, shapes and weights of the individual accidental containers are surely 'deductible,' in fact they must be so eliminable if a fair count in number of pounds of sugar is to be obtained. Galileo thought that in a similar manner the material idiosyncracies of solid bodies can be 'deducted' in considering their geometrical form and their geometrical relationships to other bodies. But what warrants this 'deduction' in the case of applied geometry? Surely the case of applying geometry is not at all that close to the case of counting up pounds of sugar. Counting quantities is one thing, applying geometrical theorems another. The contingent circumstances of the containers of the sugar are surely eliminable in an attempt to find out how much sugar is on hand. All that we need to be sure of is that each container is indeed a container of sugar, and the weights of each container. But how can we eliminate the fact (which Simplicio stresses) that in applying geometry the objects in the world are *never like* the objects talked about in geometrical systems? For purposes of counting quantities of goods in containers the individual containers are not epistemological hindrances of any kind; in the case of attempting to apply geometries to the world, the material hindrances associated with actual physical objects give rise to grave and enormous philosophical difficulties. Simplicio remarked that material objects are all porous; he might also have pointed out that they all differ in chemical composition and in a host of other ways. So how do we read away these many special material perturbations? One way is to employ exactly that physics that embodies the geometry in question. Using this geometry, we get a suitable physical interpretation for it by employing it in distinguishing different volumes, densities and other varying physical properties of the objects in question. Moving in this way prejudices the outcome in favor of the preferred geometry before any experiments or observations are made. Under such circumstances DMH would not prove any kind of case against the Aristotelians, it would simply function as a Galilean dogma. And DMH would likewise count as an anticipatory philosophy of science later championed in great detail by people like Poincaré, Duhem and Einstein.

However, I am by no means suggesting that Galileo's philosophy of geometry historically 'anticipates' the findings of the three-named 'conventionalists.' Certainly he did not seem to share Duhem's quite sophisticated thesis about the in principle non-falsifiability of single physical hypotheses. It is true that Galileo, like Poincaré and Einstein, was arguing for the retention of a specific (in this case as in theirs Euclidian) geometry, but then, unlike Poincaré and Einstein, Galileo knew no other geometries. Indeed, it was enough in his day to get opponents to see that any geometry at all could truly describe the world. However, there are certain similarities between what Galileo was trying to do and, for example, the position of Einstein with respect to geometry. His suggested argument for DMH thus raises issues similar to those raised by Einstein's position on geometry. The work of Adolf Grünbaum has shown with great clarity the range of problems generated by accepting various more restricted forms of DMH. The following statement of Einstein's view of the relationship between physics and geometry seems to me quite relevant in assessing the kind of position that Galileo was espousing when he introduced the DMH tactic:

In opposition to the Carnap-Reichenbach conception, Einstein maintains that no hypothesis of physical geometry is *separately* falsifiable, i.e., in isolation from the remainder of physics, even though all of the terms in the vocabulary of the geometrical theory, including the term 'congruent' for line segments and angles, have been given a specific physical interpretation. And the substance of his argument is briefly the following: In order to follow the practice of ordinary physics and use rigid solid rods as the physical standard of congruence in the determination of the geometry, it is essential to make computational allowances for the thermal, elastic, electromagnetic, and other deformations exhibited by solid rods. The introduction of these corrections is an essential part of the logic of testing a physical geometry. For the presence of inhomogeneous thermal and other such influences issues in a dependence of the coincidence behavior of transported solid rods on the latter's *chemical composition*, whereas physical geometry is conceived as the system of metric relations exhibited by transported solid bodies independently of their particular chemical composition. The demand for the computational *elimination* of such substance-specific distortions as a prerequisite to the experimental determination of the geometry has a thermodynamic counterpart; the requirement of a means for measuring temperature which does not yield the discordant results produced by expansion thermometers at other than fixed points when different thermometric substances are employed. This thermometric need is fulfilled successfully by Kelvin's thermodynamic scale of temperature.

But Einstein argues that the geometry itself can never be accessible to experimental falsification in isolation from those other laws of physics which enter into the calculation of the corrections compensating for the distortions of the rod. And from this he then concludes that you can always preserve any geometry you like by suitable adjustments in the associated correctional physical laws. (Grünbaum, 1973, pp. 131–32.)[7]

Grünbaum's summary of Einstein's position might be taken as a capsule view of more recent ways of worrying about Galileo's problems. It is especially interesting to note the observation that there is an epistemological link between computational elimination of physical distortions and the need for a standardizing thermodynamic scale of temperature. Carried back in time to the 17th Century, Grünbaum appears to be presenting a more detailed and physically sophisticated form of Galileo's quest for objective conditions of experimentation, conditions that would have to be up to the task of specifying how the geometry can be tested, both by the elimination of physical deformations and by the elimination of the influence of non-standardized factors which depend upon the so-called secondary qualities (note the analogy between the variability and idiosyncracy of secondary qualities and the unreliability of expansion thermometers when employed at random to measure temperatures of different thermometric substances). But one would not want to carry this exercise in anachronistic history too far. (I myself would be prepared to carry it much farther, but then I am sympathetic to the notion that the history of physics is both scientifically and philosophically continuous, a view that is badly out of fashion today. I would like to invite my readers, however, at least to ponder the connections between the long quote from Grünbaum and the major points I am trying to make in this paper.)

One thing that emerges from a comparison of Grünbaum's statement of Einstein's position on geometry and Galileo's DMH is that Galileo's use of DMH does not seem to commit him to an espousal of something like the Duhem-Einstein non-falsifiability thesis respecting geometry. That thesis requires that we accept that confronted with apparently falsifying evidence the physicist is always at liberty to change other non-geometrical parts of the physical theory.[8] Thus any specific geometry we wish can be retained in the face of recalcitrant data. Galileo's DMH is not equivalent to this form of the Duhem-Einstein thesis. It is both more general and more perplexing. Galileo is not arguing that we ought to be able to retain any geometry that we wish in the face of what appears to be falsifying evidence (recall again that he did not have available to him a knowledge of alternative geometries), rather, he is arguing that nature is geometrical essentially, and hence that any evidence that geometry does not *apply* to the world *would have to be discarded* by appeal to both DMH and specific counterfactuals like (C). Thus the argument that nature is essentially mathematical (geometrical) is that all of the theorems of geometry in physically interpreted form ((C) for example) are true counterfactuals because their antecedents will never be realized in the domain of solid objects. Thus Galileo's position has to do not with preserving a geometry in the face of falsifying evidence, but rather with trying to show that mathematics must be used in science to secure the integrity of experiments precisely because the elements in the physical world *as perceived* never satisfy the theorems of mathematics. The connection between Galileo's tactics in the *Dialogue* and his distinction between the two kinds of qualities is thus seen to be very close. Clearly the conclusion of the *Dialogue* argument that geometry is the preferred way of talking about physical objects is quite gratuitous unless we have some way of *distrusting* information about physical objects that is received by means of direct perception. (C) – and the same holds for all other physically interpreted theorems of geometry – is true precisely because no sense experiences will ever show that the antecedent holds. Unfortunately the argument for this tactical move is not given anywhere in the *Dialogue.*[9]

Galileo's position is certainly paradoxical. We are invited to believe that geometry is the most acceptable way of talking about physical objects *just because* physically interpreted geometrical theorems do not apply to perceived physical objects. As I noted earlier, had Simplicio been a bit more sharp he might have suggested that all that Galileo has shown is that physical geometry is *false*. Alternatively he might have suggested that Salviati needed more argument to sustain the conclusion that geometry is the supreme form of scientific knowing. We are also invited to accept DMH on a ridiculous analogy with weighing pounds of sugar. What is going on? Again I think the key to removing perplexity requires a shift in emphasis: although Galileo's strategy is ontologically very rich, he uses that very richness in a misleading way. For it is not an ontology *per se* that mainly engages the great talents of Galileo, it is an ontology in the service of introducing a new methodology that provided a central place for mathematics in the experimental manipulation of ordinary sensory experience. Thus it should not seem too strange that Galileo does not make good philosophically on the claim that mathematics is the language of nature; he does present an appealing case for shifting our methodological and hence epistemological emphasis. The tactical moves sustain the strategy of the methodology: the world of matter is not the world of physical objects as perceived, it is the world of physical objects as determined by experiments.

We have seen that Galileo's tactical moves on behalf of establishing the supremacy of mathematical forms of physical explanation rest upon combining (1) acceptance of DMH, (2) the thesis that all physically interpreted theorems of geometry (e.g., (C)) are true counterfactuals, and (3) the distinction between primary and secondary qualities. In have made some attempt to make the combination of (1)–(3) into an argument that is at least plausible; but that argument is simply unsatisfactory. (1) is only acceptable if we know a priori that (2) is true and also that (3) is true. Galileo offers no argument for (3) and the quasi-argument for (2) seems to presuppose that he will not give up geometrical forms of explanation under any circumstances. Galileo's Aristotelian opponents are perfectly at liberty to interpret his conclusions in a quite different way. The simplest move for them to make is to deny (3), which of course they did. We seem again to have to agree with Feyerabend. Galileo made his case not so much by means of argument as by means of successful propaganda. Part of the ultimate success of the propaganda rests upon the substantive support given to his philosophical programme by the lasting results of his positive science.

The general position of my paper might be given a final boost by looking briefly at one of Galileo's positive scientific achievements, his free fall law (acceleration-rate law). The law is simple and well-known: all physical bodies in a state of free fall (regardless of their differences in weight, the property the Aristotelians had fastened attention upon) accelerate at the same rate in the direction of the surface of the earth. The way in which Galileo arrived at this law is a classic case of 'deducting the material hindrances.' His own experiments with rolling balls down inclined planes had shown him that according to normal commonsense observation (Aristotelian observation?) the law of equal acceleration does not hold. The Aristotelians had concluded that the acceleration of a falling body depends upon its weight and also upon the medium through which it is moving. Galileo's insight consists in denying both Aristotelian points. But with what justification? The medium through which an object is moving creates impediments of several kinds: resistance, specific weight, shape of the moving body, and contact between the surface of the moving body and the fluid medium.[10] Because of the effects of these and other 'hindrances' the proposed law should have been taken to be false. But Galileo's unwavering confidence in deductive inference and in mathematics generally convinced him that the hindrances were indeed only that, and that they should be shown to be inessential, rather than taken as disconfirmations of the law. Here his way of proceeding more nearly approximates the Duhem-Einstein form of arguing for preferred formalisms. Again I will defer to contemporary *noiseologies* (doctrines full of 'sound and fury' usually 'signifying nothing'). We will therefore have to construe Galileo's odd way of defending his 'law' as being merely a case of DMH. At least this way of understanding him is 'safe.'

Galileo observed that as the medium through which a physical object moved became less dense the more the movement of the object conformed to the law of free fall. This means that given two objects of different weights, and a medium through which they are moving whose density approximates zero, the two objects will accelerate at the same rate. Characteristically, Galileo had no experimental evidence for this conclusion, indeed, he thought that the required experiment could not be performed. The final conclusion that he reached – treating the medium of movement as an impediment rather than as a disconfirmation – was that in a vacuum all bodies regardless of weight would accelerate at the same rate – *at exactly the same rate*. And he believed this even though he thought that it is perhaps impossible to produce a vacuum! Thus his proposed law is both daring and outrageous. To believe it – indeed even to entertain it – requires that we suspend trust in what we ordinarily observe. We are invited to a magic show, and the magician is not even confident that he can bring off the trick! Of course we now know that in an evacuated vacuum tube pennies and feathers fall side-by-side. We know that Galileo was right on the basis of experiments that he did not himself perform, and thought impossible. Once again we are requested to accept the truth of a set of counterfactuals without compelling evidence.

Nevertheless there can be little doubt that Galileo hit upon exactly the propositions that had to be believed in order to have confidence in mathematics as the language of science. His genious lay in part in being able to envision logical possibilities (and in the case of the free fall law, what he thought of as experimental *impossibilities*) that required non-standard and frequently unavailable observations for their substantiation. However, there is nothing in his tactical behaviour that constrains belief. In the end one is still able to be open: the material hindrances can be taken as essential parts of the laws, as disconfirmations of the laws, or just as hindrances – eliminable deformations. Shea has put something like this point admirably:

The scientific revolution has taught us that by compelling nature to do, so to speak, what it does not naturally do, new truths about the structure of the universe are often disclosed. For us, this assertion is supported by three hundred years of experimental science. For Galileo's contemporaries, and indeed for Galileo himself, who even doubted the experimental possibility of producing a vacuum, it was a leap of faith. (Shea, 1972, p. 162.)

So Galileo's confidence in mathematics and in the efficacy of experimentation must be taken as a philosophically unjustified confidence. In part the confidence is justified *ex post facto* by the positive results of physics built upon his expectations, and such success was one of the outcomes that Galileo trusted. But his mathematical realism had to wait for the philosophical justification he himself had failed to provide.

We have seen that Galileo's confidence in mathematics culminated in a rush of bravado. His belief that in knowing geometrical truths human understanding is identical to God's understanding remained unsupported. The distinction between primary and secondary qualities needed for his attack upon Aristotelianism was undefended, and his argument that mathematics applies to the world was more a metaphysical faith than a philosophically established conclusion. He seems to have concluded that if the world does not conform to truths of mathematics, so much the worse for the world. Nevertheless, Galileo's philosophical contribution was in the long run very great indeed. Like all great revolutionaries, he proposed an alternative that eventually became irresistible, partly because of the success of his kind of science and partly because his philosophical daring pointed out the specific problems attendant upon an Aristotelian empiricism. Although he was unable – or was uninterested in – working out the details of his system, it is probable that what he fostered in fact does involve a coherent set of propositions that could be taken to be something like 'Galileo's philosophy.'

Among those propositions are the following:

(1) Science is not about the things that naked-eye observations tell us about, it is about experimental possibilities expressed in mathematical terms.

(2) At a certain regulative level – a level at which methodological considerations outweigh epistemological ones – experimentation is not an attempt to confirm theory by repetition; it is rather a way of envisioning theoretical possibilities, where those possibilities always depend upon viewing reality as sets of mathematical

properties.[11]

(3) Matter is not available to ordinary perception, it is geometry interpreted physically.

All three propositions depend upon our willingness to reject commonsense observation as the basis of science. All three also depend upon our resolve to entertain possibilities that completely outstrip our observable possibilities. These considerations entail that science must be prepared to deal with *contrived situations*. An experiment, after all, is precisely the creation of non-normal (by commonsense standards) or artificial situations. The final conclusion is clear: *scientific* experience – the kind of experience we must be able to have in order to determine the mathematical possibilities to be true or false – is not at all the kind of experience that Aristotle and his followers thought was basic. It is rather that kind of experience that is publically available to all, given that the observers have the requisite wit, training and intelligence.

Thus however we measure Galileo's specific contributions to science – and these were many – it remains to appreciate that his philosophical programme was propogandistic and consequently logically unsettled. His importance is therefore misunderstood by historians of both science and philosophy. The conceptual significance of his systematic polemics against the Aristotelian stereotype may have been vastly over-rated. I think, however, that one matter is clear: had Galileo never existed, a large number of subsequent scientists would have had to invent him, beginning, perhaps, with Newton.

The University of Western Ontario

NOTES

[1] Like many people who write on Galileo I am not a Galileo specialist. I am aware that many of his contemporaries shared views similar to his on matters both philosophical and scientific. I am also aware – but in this paper will ignore – that a fairly substantial tradition of experimentation predated Galileo's work. These historical concessions in no way militate against the interpretation of Galileo that I will offer.

[2] Even if what follows confirms Feyerabend's views of Galileo, there is no cause for alarm. What Galileo introduces on behalf of his programme for mathematizing scientific experience is only a sketch (negatively expressed – a caricature). Descartes, Newton and Leibniz do a better metaphysical job. Only Kant, however, sees the true limits of the philosophical task, and sets himself the required problems. There is philosophy as the

introduction of bold, seemingly inappropriate ideas. There is philosophy as the analysis of the final success of such ideas. More than anything else it is the improbable link between Galileo and Kant that makes Galileo's bravado both tolerable and interesting – at least to those of us who see his ideas as having some lasting effect upon modern *philosophical* norms. It is not of minor importance that Kant recognized this achievement of Galileo in his Preface to the second edition of *Critique of Pure Reason*. I will not in this note undertake to write the additional paper prerequisite to an understanding of Kant's insight.

[3] Galileo seems to have had something like this view of the two kinds of qualities in mind for a long time. He writes in 'Letters on Sunspots. . .', p. 124: "Hence I should infer that although it may be vain to seek to determine the true substance of the sunspots, still it does not follow that we cannot know some properties of them, such as their location, motion, shape, size, opacity, mutability, generation, and dissolution. These in turn may become the means by which we shall be able to philosophize better about other and more controversial qualities of natural substances. And finally by elevating us to the ultimate end of our labors, which is the love of the divine Artificer, this will keep us steadfast in the hope that we shall learn every other truth in Him, the source of all light and verity." [Galileo here writes partially in the Aristotelian idiom typical of his approach in the 'Letter.' But the fact is clear that he is already prepared to identify knowledge of some properties (geometrical and quantitative ones) with divine knowledge.]

[4] Compare Galileo's form of the distinction with what Descartes has to say about the ball of wax in the second 'Meditation.' Inanities of standard history of philosophy aside, both Galileo and Descartes seem to have had something of great importance in mind when they introduced the distinction. But clearly neither of them thought of the distinction as underwriting a specific form of empiricism or rationalism (but then of course neither of them had read Hegel's history of philosophy!). Drake's note is in Drake, 1957, p. 274.

[5] As we shall see there is a certain simplistic assumption underlying Galileo's distinction, the assumption that sensory and what appear to be essentially private qualities cannot be made accessible to measurement. Half a programme is better than no programme at all. Nevertheless, there are many hints in his writings that Galileo at least entertained the notion of rendering secondary qualities fit for scientific measurement.

[6] I have put this point somewhat unfairly and perhaps – so far as Galileo scholars are concerned – flippantly. But not much is at stake. Salviati had already argued that if a body is projected along a tangent struck anywhere on the surface of the earth it would not be released from the surface of the earth because of the acuteness of the angle involved. Shea (1972, pp. 140–42) has shown that Salviati failed to take into account the centrifugal force of the moving earth. Given sufficient speed of centrifugal movement, the earth would throw off at least some bodies; and this is all that Simplicio needs for purposes of ridicule. Salviati's real point comes next.

[7] Chapter 4 of this book provides an elegant discussion of problems connected with conventionalism in geometry that in a sense 'update' Galileo's implicit concerns in the *Dialogue*. Of equal importance is the general treatment of falsifiability in science in Chapter 17. Einstein's fullest statement of his position on geometry is in Einstein, 1949, pp. 676–79.

[8] I am here regarding the Duhem-Einstein thesis as a special case of the more general Duhem thesis. Roughly phrased, Duhem's thesis points out that in the case of well

articulated physical theories, negative experimental results cannot logically falsify the entire physical theory at issue, but only some (logically unspecifiable) aspect of that theory. This entails that the scientist has a choice – not to be made on empirical grounds – of which parts of the theory he will save. Galileo's DMH is logically much stronger than the Duhem thesis: it asserts that *no matter what* a certain philosophical theory, complete with its ontology, will hold. The philosophical theory will of course be accepted for reasons far transcending the results of positive experimentation; indeed, the reasons will *shape* the very nature of experimentation.

[9] Galileo's 'The Assayer' appeared in 1623, his *Dialogue* in 1632. In view of the fact that the *Dialogue* presents a massive frontal attack upon the stereotypical Aristotelian physics, including its reliance upon direct reports of perception, I think it is fair to assume that Galileo's distinction in 'The Assayer' is presupposed in the later work. See note 3 above for suggestions of Galileo's early indications of acceptance of the distinction.

[10] Shea (1972, pp. 159–63) contains an excellent brief discussion of Galileo's free fall law. My own account owes much to Shea's treatment of the topic.

[11] Shea (1972, pp. 39–40) provides an insightful discussion of Galileo's use of experiments as regulative principles. He writes: "It would seem, therefore, that experiments are not essential for Galileo in the sense that their mere mechanical repetition can produce a theory. Rather they are important inasmuch as they play a discriminatory role in the selection of the set of principles that will be used as the basis of a physical interpretation of nature. This means that framing exact hypotheses is only the first step in science. The second one is deriving practical conclusions from them and devising well-chosen experiments to test them. It is one of Galileo's great contributions to the development of scientific method that he clearly recognised the necessity of isolating the true cause by creating artificial conditions where one element is varied at a time." My own fuller treatment of this and related themes is in Butts (forthcoming).

BIBLIOGRAPHY

Burtt, E. A., 1932, *The Metaphysical Foundations of Modern Physical Science*, revised ed., N. Y., Garden City, 1954.

Butts, Robert E., forthcoming, 'Experience and Experiment as Regulative Principles in Methodology', in *Boston Studies in the Philosophy of Science*.

Drake, Stillman (transl.), 1957, *Discoveries and Opinions of Galileo*, Garden City, N.Y.

Einstein, Albert, 1949, 'Reply to Criticisms', in P. A. Schilpp (ed.), *Albert Einstein: Philosopher-Scientist*, New york.

Feyerabend, P. K., 1970, 'Problems of Empiricism II', in *The Nature and Function of Scientific Theories*, (ed. by R. Colodny), Pittsburgh.

Galileo Galilei, 1613, 'History and Demonstrations concerning Sunspots and their Phenomena', in Drake (1957).

Galileo Galilei, 1623, 'The Assayer', in Drake (1957).

Galileo Galilei, 1632, *Dialogue Concerning the Two Chief World Systems* (transl. by Stillman Drake), Berkeley and Los Angeles.

Grünbaum, Adolf, 1973, *Philosophical Problems of Space and Time*, 2nd enlarged ed., Reidel, Dordrecht.

Koyré, Alexandre, 1939, *Etudes Galiléennes*, Paris.
Koyré, Alexandre, 1943, 'Galileo and Plato', *Journal of the History of Ideas* 4, 400–428.
Shea, William R., 1972, *Galileo's Intellectual Revolution*, New York.

WILLIAM A. WALLACE

GALILEO GALILEI AND THE *DOCTORES PARISIENSES**

The title of this study translates that of a note published by Antonio Favaro in 1918 in the transactions of the Accademia dei Lincei, wherein he presented his considered opinion of the value of Pierre Duhem's researches into the 'Parisian precursors of Galileo.'[1] Earlier, in 1916, only a few years after the appearance of Duhem's three-volume *Études sur Léonard de Vinci*, Favaro had reviewed the work in *Scientia* and had expressed some reservations about the thesis there advanced, which advocated a strong bond of continuity between medieval and modern science.[2] In the 1918 note he returned to this topic and developed a number of arguments against the continuity thesis, some of which are strikingly similar to those offered in present-day debates. Since Favaro, as the editor of the National Edition of Galileo's works, had a superlative knowledge of Galileo's manuscripts – one that remains unequalled in extent and in detail to the present day – it will be profitable to review his arguments and evaluate them in the light of recent researches into the manuscript sources of Galileo's early notebooks. Such is the intent of this essay.

I. FAVARO'S CRITIQUE OF THE DUHEM THESIS

The purpose of Favaro's 1916 review, as shown by its title, was to question whether Leonardo da Vinci really had exerted an influence on Galileo and his school. To answer this Favaro focussed on the concluding portion of Duhem's final tome, wherein Duhem had cited the first volume of Favaro's National Edition containing the previously unedited text of Galileo's Pisan notebooks. Duhem had noted in Favaro's transcription of the manuscript containing the notes the mention of Heytesbury and the 'Calculator,' the discussion of degrees of qualities in which Galileo uses the expression *uniformiter difformis*, and explicit references to the *Doctores Parisienses*.[3] For Duhem these citations, and particularly the last, clinched the long argument he had been developing throughout his three volumes. Galileo himself had given clear indication of his Parisian precursors: they were Jean Buridan, Albert of Saxony, and Themo

R. E. Butts and J. C. Pitt (eds.), New Perspectives on Galileo, 87–138.

Judaeus – and here Duhem even ventured the volume from which Galileo had extracted his information, the collection of George Lokert, published at Paris in 1516 and again in 1518.[4] So the continuity between medieval and modern science was there for all to see, spelled out in Galileo's youthful handwriting.

Favaro, however, was not convinced. He saw Duhem's effort as partial justification for a general thesis to which he himself did not subscribe, namely, as he put it,

> that the history of science shows scientific development to be subject to laws of continuity, that great discoveries are almost always the fruits of a slow and involved preparation worked out over centuries, that. . .[they] result from the accumulated efforts of a crowd of obscure investigators. . .[5]

For Favaro such a thesis was too disparaging to the truly great men, to Newton, Descartes, his own Galileo; it accorded insufficient credit to these fathers of modern science.

But Favaro's refutation of Duhem was not argued at this level. Instead, proper historian that he was, Favaro turned to the manuscript itself. Admittedly it was in Galileo's handwriting, and Favaro was the first to concede that, but this proved only that it was the work of Galileo's hand, not necessarily the work of his head. The notes were extremely neat, spaces were occasionally left to be filled in later, some of the later additions did not fit and spilled over into the margins, there were signs of copying – phrases crossed out and reappearing a line or two later – all indications that this was not an original composition. Moreover, the notes were very sophisticated, they had a magisterial air about them, quoting authorities and opinions extensively – in a word, they manifested considerably more learning than one would expect of the young Galileo.[6] And 'young' Galileo truly was when these notes were written. Internal evidence could serve to date them, as Favaro had already indicated in his introduction to the edited text. "Without any doubt," he had written, "Galileo wrote these notes during the year 1584."[7] That is why Favaro himself had entitled them *Juvenilia*, or youthful writings, perhaps stretching the term a bit, for his 'youth' was by then twenty years of age. Though in Galileo's hand, therefore, they were but his work as a student at the University of Pisa, probably copied from one of his professors there – a likely candidate would be Francesco Buonamici, who later published a ponderous tome of over a thousand folio pages discussing just these problems.[8] So one ought not make too much of references to the *Doctores*

Parisienses. The *Juvenilia* are not Galileo's expression of his own thought, they do not represent his work with original sources, not even with Duhem's 1516 edition of Parisian writings. They are materials copied at second or third hand, the scholastic exercises of a reluctant scholar who probably had little taste for their contents, and indeed would soon repudiate the Aristotelianism their writing seemed to imply.[9]

Favaro's refutation of the Duhem thesis, so stated, is difficult to counter, and it is not surprising that it has remained virtually unchallenged for over fifty years. This in spite of two facts that go against Favaro's account, to wit: (1) there is no mention of the *Doctores Parisienses* in Buonamici's text, or indeed, in the writings of any of Galileo's professors who taught at Pisa around 1584, nor do any of them discuss the precise matters treated in the notes, and (2) the curator who, considerably before Favaro, first collected Galileo's Pisan manuscripts and bound them in their present form had made the cryptic notation: "The examination of Aristotle's work *De caelo* made by Galileo around the year 1590."[10] Now in 1590, as we know, Galileo was by no accounting a youth; he was 26 years of age, already teaching mathematics and astronomy at the University of Pisa. So Favaro's critique leaves two questions unanswered: (1) if Galileo copied the notes, from whom did he copy them, particularly the references to the *Doctores Parisienses*; and (2) must they be the notes of an unappreciative student, even one twenty years of age, or could they be the later work of an aspiring professor who understood the arguments contained in them and possibly made them his own? Depending on how we answer these questions we may have to revise Favaro's judgment on the value of the notes, and in particular their bearing on his critique of the Duhem thesis.

II. MS GAL 46: CONTENTS AND CITATIONS

Before attempting a reply it will be helpful to review the contents of the notebooks that have been labelled for so long as Galileo's *Juvenilia*. Actually they are made up of two sets of Latin notes, the first containing questions relating to Aristotle's *De caelo et mundo* and the second containing questions relating to Aristotle's *De generatione et corruptione*. These questions are dealt with in a stylized scholastic manner, first listing various opinions that have been held, then responding with a series of conclusions and arguments in their support, and finally solving difficulties that have been raised in the opinions of those who hold the contrary. Each of Galileo's sets of notes contains twelve questions that follow this general pattern, but with some variations. The first set begins with two introductory questions, then has four questions that make up its *Tractatio de mundo*, followed by six questions of a *Tractatio de caelo*, the last question of which is incomplete. The second set of notes, like the first, is incomplete at the end, but it is also incomplete at the beginning. It comprises three questions from a *Tractatus de alteratione* that discuss intensive changes in qualities, and nine questions of a *Tractatus de elementis*, five of which treat the elements of which the universe is composed and the remaining four the primary qualities usually associated with these elements.[11] Apart from the fact that both sets of notes are patently incomplete, there are numerous internal indications that they either are, or were intended to be, parts of a complete course in natural philosophy that would begin with questions on all the books of Aristotle's *Physics*, go through the *De caelo* and the *De generatione*, and terminate with a series of questions on the *Meteorology*.[12] Whether this complete set of notes was actually written by Galileo and subsequently lost, or whether what has been preserved constitutes the whole of his writing on these subjects, must remain problematical.

Unlike much of Galileo's later composition, these notes are replete with citations – a fortunate circumstance, for such citations are frequently of help in determining the provenance of manuscripts of that type. In this particular manuscript Galileo cites 147 authors, with short titles of many of their works, as authorities for the opinions and arguments he adduces. Many of these are writers of classical antiquity, the authors of primary sources such as Plato, Aristotle, and Ptolemy. But a goodly number of medieval and Renaissance authors are mentioned also, and, based on frequency of citation, it was to these later sources that Galileo seems to have had recourse more generally. Indeed, it is instructive to list the principal authors he cites in the order of frequency of citation, as shown in Table I, since this reveals some curious facts. Note in this list, for example, that the author who is most frequently referenced is one of Aristotle's medieval commentators, Thomas Aquinas, who is mentioned 43 times. After him comes Averroës, 38 times, and then a group of Dominicans whom Galileo refers to as 'the Thomists' (*Tomistae*), whose citations singly and collectively total 29.[13] After them come classical

TABLE I

A Selection of Authors Cited in MS Gal 46
Including All the Most Recent Authors and
Listed in Order of Frequency of Citation

Aquinas		43	Pomponatius	4
Averroes		38	Taurus	4
Thomistae	4	29		
Capreolus	7		Biel	3
Caietanus	6		Buccaferreus	3
Soncinas	4		Cardanus	3
Ferrariensis	3		Contarenus	3
Hervaeus	2		Diogenes Laertius	3
Soto	2		Nobili	3
Javelli	1		Pererius	3
Simplicius		28	Pliny	3
			Pythagoras	3
Plato		20	Scaliger	3
Ptolemy		20		
Philoponus		19	Alfraganus	2
Albertus Magnus		17	Algazel	2
Alexander Aphrodisias		16	Alpetragius	2
			Antonius Andreae	2
Galen		15	Avempace	2
Aegidius Romanus		12	Avicebron	2
Anaxagoras		11	Burley	2
Avicenna		10	Copernicus	2
Duns Scotus		10	Doctores Parisienses	2
Plutarch		10	Gaetano da Thiene	2
			Iamblichus	2
Durandus		9	John Canonicus	2
Regiomontanus		9	Nifo	2
Augustinus		8		
Empedocles		8	Albumasar	1
Achillini		7	Aristarchus	1
Jandunus		7	Balduinus	1
Proclus		7	Carpentarius	1
Themistius		7	Cartarius	1
Ockham		6	Clavius	1
Porphyry		6	Crinitus	1
Zimara		6	Heytesbury	1
			Lychetus	1
Albategni		5	Marlianus	1
Bonaventura		5	Nominales	1

Democritus	5	Pavesius	1
Marsilius of Inghen	5	Philo	1
Pietro d'Abano	5	Philolaus	1
		Puerbach	1
Alfonsus	4	Sixtus Senensis	1
Henry of Ghent	4	Strabo	1
Hippocrates	4	Swineshead	1
John of Sacrobosco	4	Taiapetra	1
Mirandulanus	4	Valeriola	1
Paul of Venice	4	Vallesius	1
Plotinus	4		

commentators, such as Simplicius and Philoponus, Plato and Ptolemy with 20 citations apiece, Galen with 15, Scotus with 10, Ockham with 6, Pomponazzi with only 4, the *Doctores Parisienses* with 2, and over a score of other authors with only 1 apiece.

In looking over this list, one might wonder whether printed editions of any of these authors were available to Galileo, for if these could be identified and located, they might provide a clue to dating the notes, or even turn out to be the source or sources from which the notebooks were compiled. Pursuing this lead, one finds that, of the most recent titles cited by Galileo, 27 were works printed in the sixteenth century, all of which are listed in Table II. Some of

TABLE II

Sixteenth-Century Printed Sources Cited in MS Gal 46*

Balduinus, Hieronymus, *Quaesita. . .et logica et naturalia*, Venice 1563
Buccaferreus, Ludovicus, *In libros de generatione*, Venice 1571
Caietanus, Thomas de Vio, *In summam theologicam*, Lyons 1562
Cardanus, Hieronymus, *De subtilitate*, Nuremberg 1550
Carpentarius, Jacobus, *Descriptio universae naturae*, Paris 1560
Cartarius, Joannes Ludovicus, *Lectiones super Aristotelis proemio in libris de physico auditu*, Perugia 1572
Clavius, Christophorus, *In sphaeram Ioannis de Sacrobosco*, Rome 1581†
Contarenus, Gasparus, *De elementis*, Paris 1548
Copernicus, Nicolaus, *De revolutionibus*, Nuremberg 1543
Crinitus, Petrus, *De honesta disciplina*, Lyons 1554
Ferrariensis, Franciscus Sylvester, *In libros physicorum*, Venice 1573
Ferrariensis, Franciscus Sylvester, *In summam contra gentiles*, Paris 1552
Javellus, Chrysostomus, *Totius. . .philosophiae compendium*, Venice 1555

Lychetus, Franciscus, *In sententiarum libros Scoti*, Venice 1520
Mirandulanus, Bernardus Antonius, *Eversionis singularis certaminis libri*, Basel [1562?]
Niphus, Augustinus, *In libros de generatione*, Venice 1526
Nobilius, Flaminius, *De generatione*, Lucca 1567
Pavesius, Joannes Jacobus, *De accretione*, Venice 1566
Pererius, Benedictus, *De communibus omnium rerum naturalium principiis*, Rome 1576††
Pomponatius, Petrus, *De reactione*, Bologna 1514
Scaliger, Julius Caesar, *Exercitationes de subtilitate. . .ad Cardanum*, Paris 1557
Sixtus Senensis, *Bibliotheca sancta*, Venice 1566
Sotus, Dominicus, *Quaestiones in physicam Aristotelis*, Salamanca 1551
Taiapetra, Hieronymus, *Summa divinarum et naturalium quaestionum*, Venice 1506
Valleriola, Franciscus, *Commentarium in libros Galeni*, Venice 1548
Vallesius, Franciscus, *Controversiarum medicarum et philosophicarum libri*, Alcala 1556
Zimara, Marcus Antonius, *Tabula delucidationum in dictis Aristotelis et Averrois*, Venice 1537

---

* The list cites the earliest known printed edition, except for the works of Clavius and Pererius.
† See note 20, *infra*.
†† See note 22, *infra*.

these are by authors well known to historians of science, such as Cardanus, whose *De subtilitate* was printed at Nuremberg in 1550, and Copernicus, whose *De revolutionibus* was printed there in 1543. Similarly well known are the works of Nifo, Pomponazzi, and Zimara, all from earlier decades of the sixteenth century. Other works are less well known, for example, Nobili's *De generatione*, which was printed at Lucca in 1567. But only one book on the list, as it turns out, was published in the 1580's. That is Christopher Clavius's *In sphaeram Ioannis de Sacrobosco commentarius*, the second edition of which was printed at Rome in 1581. Clavius, of course, was the famous Jesuit, 'the Euclid of the sixteenth century,' who taught mathematics and astronomy at the Collegio Romano, the influential Jesuit university in Rome. And, by coincidence, the next most recently published work on the list is that of another Jesuit professor at the Collegio Romano, Benedictus Pererius, an edition of whose *De communibus omnium rerum naturalium principiis* was published in Rome also, but five years earlier, in 1576.

Moreover, if one looks at a tabulation of the places and dates of earliest imprints of these volumes, as shown in Table III, some interesting results emerge. Most of the books, as might be expected, were printed at Venice, and

TABLE III

Sixteenth-Century Printed Sources Cited in MS Gal 46

Places of earliest imprints	
Venice	11
Paris	4
Rome	2
Lyons	2
Nuremberg	2
Bologna	1
Perugia	1
Lucca	1
Basel	1
Salamanca	1
Alcala	1

Dates of earliest imprints	
1580's	1
1570's	4
1560's	7
1550's	7
1540's	3
1530's	1
1520's	2
1510's	1
1500's	1

Italian publishing houses predominate, although there are some French, German, and Spanish also. With regard to the dates of imprint, books published in the 1550's and 1560's are in the majority, with some from the 1570's and one from the 1580's. If Galileo composed these notes from printed sources, therefore, he must have had access to a good library with a wide variety of works that were kept fairly up to date. More important, if he composed them from printed sources, he must have done so *after* 1581 – a point to be recalled later.

Since Clavius and Pererius are the most recent authors cited in MS Cal 46, it would appear worthwhile to consult their works for any evidences of copying that would tie Galileo to these sources. During the past few years this has been done by Alistair Crombie, Adriano Carugo, and the writer, with results that are extremely gratifying. Crombie has shown that practically all of the astronomy contained in MS Gal 46, in fact, could have been taken almost verbatim from Clavius's commentary on the *Sphere* of Sacrobosco, and Carugo, on Crombie's report, has shown that other sections that treat of the composition of the heavens and the intension and remission of forms could have derived from Pererius's *De communibus omnium rerum naturalium principiis.*[14] Taken together, by my estimate the materials contained in these two Jesuit authors can account for about 15% of the entire contents of MS Gal 46. Moreover, though it is possible that Galileo could have copied the notes from some intermediate author who in turn based them on Clavius and Pererius, there is *prima facie* evidence that Galileo himself used these two authors when composing the notes. And whether he did so or not, Favaro's suspicion is partially confirmed by these discoveries: Galileo clearly did not compose his notes from primary sources alone, although he supplies references to such sources. Rather he freely utilized the citations found in more recent works such as those of Clavius and Pererius, or of some even later writer whose excerpts from these authors were somehow made available to him.

III. MS GAL 46: CLAVIUS AND PERERIUS

Since the relationships between Galileo's notes and the printed works of these two Jesuits will assume considerable importance in what follows, it will be well to review here the *prima facie* evidence that supports the view that Galileo had access to these two books and actually copied from them, rather than from some intermediate source.

Looking first at Crombie's evidence, the portion of Galileo's manuscript that seems to be based on Clavius consists of 10 folios of the 97 that make up the two sets of notes. They constitute the first two questions of the *Tractatio de caelo* already referred to, which treat respectively of the number and the ordering of the celestial orbs. Galileo begins the initial question of this treatise by listing opinions, the first of which he cites as follows:

> The first opinion was that of certain ancient philosophers, whom St. Chrysostom and some moderns follow, holding that there is only one heaven. Proof of this opinion: all of our knowledge arises from the senses; yet, when we raise our eyes to heaven we do not perceive several heavens, for the sun and the other stars appear to be in one heaven; therefore [there is only one heaven]. Nor do the heavens fall under any sense other than sight.[15]

To show the comparison of texts that suggests copying, the Latin versions are reproduced below in parallel column, with the transcription of Galileo's hand on the right and the relevant portions of Clavius from which the notes were apparently taken on the left:

CLAVIUS

In *Sphaeram*. . .

Commentum in primum caput

De numero orbium caelestium

Antiquorum[1] philosophorum nonnulli[2] unicum duntaxat caelum esse affirmabant, quos pauci[3] admodum ex recentioribus[4] imitantur[5]. . .Omnis scientia[6] nostra. . .a sensu oritur.[7] Cum igitur, quotiescunque[8] ad caelum oculos attolimus, non percipiamus[9] visu multitudinem caelorum, Sol enim[10]. . .et reliquae. . .stellae[11] in uno. . .caelo videntur[12] existere, caelumque ipsum sub nullum[13] alium sensum praeter[14] visum, cadere possit[15]. . .

GALILEO

Tractatio de caelo

Quaestio prima

An unum tantum sit caelum?

Prima opinio fuit veterum[1] quorumdam[2] philosophorum, quos secutus est[5] D. Chrysostomus et aliqui[3] recentiores,[4] sententium unicum esse caelum. Probatur haec opinio: omnis nostra cognitio[6] ortum habet[7] a sensu: sed, cum[8] attolimus oculos ad caelum, non percipimus[9] multitudinem caelorum, cum[10] sol et reliqua astra[11] in uno caelo videantur[12] existere: ergo [etc.]. Neque[13] vero sub alium[13] sensum quam[14] sub visum cadunt.[15]

Note here that, except for Galileo's reference to St. John Chrysostom, about which more will be said later, all of the information contained in Galileo's text is already in Clavius, although expressed in slightly different words. Note also that instances of word changes are shown in the parallel columns by superscript numbers: for example, the superscript 1 indicates that where Clavius has *antiquorum* for 'ancient' Galileo has the synonym *veterum*; superscript 2 shows the substitution of *quorumdam* for *nonulli*, and so on. Now it is possible to proceed in this fashion through the entire 10 folios that contain these two questions and to find places in the 1581 edition of Clavius from which practically every sentence in Galileo's notes could have been obtained.[16] Clavius, of course, has a more detailed treatment, since in addition to treating the number and the ordering of the spheres he also discusses their motions and periodicities, the trepidation of the fixed stars, and the earth's location at the center of the universe.[17] Even when restricting attention to the number and ordering of the spheres, moreover, Galileo's text summarizing the relevant passages contains roughly half the words used by Clavius to explain the same material.[18] Significantly, Galileo's text is not only abbreviated but in some places incorporates phrases and even entire clauses from Clavius's commentary. Usually, however, the sentence structure is changed, the inflection of words is altered, and synonyms are employed, as seen in the foregoing sample with the numbered superscripts. Withal the abbreviation is done skillfully, and in particular the order of Clavius's arguments is rearranged in a consistent pattern of exposition, so that the work is clearly that of a person who knows the subject matter well and is attempting to present it as synthetically as possible.

To check the stylistic characteristics of the author who worked from Clavius's text, presumably Galileo, a detailed analysis was made of all the word preferences and changes of inflection that occur in the noted 10 folios of MS Gal 46 vis-à-vis the 31 pages of Clavius's text on which they seem to be based. The system of superscript numbering shown in the above sample was continued for the entire two questions, and it was found that in all there were about 630 numbered expressions indicating conversions, i.e., that 630 of the 5650 words, or some 11%, were changed from the way they appear in Clavius. Of these total conversions, about 180 (or 35%) were mere changes of inflection, the word-stem remaining the same, whereas in the remaining 450 cases the word had been replaced by a synonym. By preparing index cards for each of the words changed, and counting the number of conversions to the word and from the word, it was possible to calculate an index of the author's stylistic and word preferences. The results of these calculations are presented in Table IV entitled 'List of Conversions,' with the 'likes' (those for which the conversions *to* the word exceed those *from* the word) shown on the left and the 'dislikes' (wherein the conversions *from* the word exceed those *to* the word) on the right. The changes are categorized according to the parts of speech and inflected forms, and only changes with a significant conversion index are included. A comparison of the 'likes' with the 'dislikes' listed in Table IV shows that the author's latinity is unsophisticated, that he usually prefers the simple word, and exhibits none of the classical variety of Clavius's excellent Latin prose. Many of the changes of inflection can be explained by the writer's attempt to abbreviate and synthesize, as in the preference for specific relative pronouns, but generally his changes in verbs, nouns, and

TABLE IV

List of Conversions

Conversion Index = no. of conversions *to* the word (+)/no. of conversions *from* the word (–)

LIKES		DISLIKES	
PARTICLES (187 total conversions):			
nam	+22/–2	enim	+0/–25
et	+17/–4	quare	+0/–7
ut	+14/–2	quoniam	+0/–6
cum	+12/–3	deinde	+0/–5
sed	+6/–0	denique	+0/–4
neque	+5/–0	unde	+0/–3
quia	+4/–0		
INFLECTED FORMS (179 total conversions):			
Change of Conjugation (98 total conversions):			
Subjunctive	+17/–0	Infinitive	+18/–22
Change of Declension (81 total conversions):			
Accusative	+26/–14	Nominative	+21/–35
VERBS (122 total conversions):			
movere	+9/–0	collocare	+0/–7
patet	+7/–0	statuere	+0/–6
constituere	+5/–0	deprehendere	+0/–5
esse	+10/–5	efficere	+0/–4
posse	+4/–0	dicere	+5/–9
debere	+4/–0	colligere	+0/–3
NOUNS (58 total conversions):			
orbis	+8/–3	caelum	+4/–7
occasum	+4/–1	occidens	+1/–4
cursus	+3/–0	circulus	+0/–3
corpus caeleste	+2/–0	stella	+2/–4
astronomus	+2/–0	sphaera	+4/–5
PRONOUNS (33 total conversions):			
qui, quae, quod	+12/–2	is, ea, id	+0/–7
ille, illa, illud	+12/–0		
hic, haec, hoc	+6/–1		
ADJECTIVES (36 total conversions):			
-- nothing significant			
VARIA (16 total conversions):			
-- nothing significant			

adjectives are so content-determined that they cannot be used as stylistic indicators. The particles, however, and particularly the conjunctions, show strong preferences for simple words like *nam*, *et*, *ut*, and *cum*, and almost equally strong rejections of slightly longer connectives such as *enim*, *quare*, *quoniam*, and *deinde*.

Since the foregoing use of conjunctions would be a likely characteristic of Galileo's somewhat simple style – his Latin by his own admission was never good[19] – a further study was made of the frequency of occurrence of the conjunctions shown in the 'likes' and 'dislikes' columns of Table IV to see how these occur throughout the entire 97 folios of MS Gal 46. The results of this analysis are shown in Table V, where the same consistent pattern of preference can be seen. The two middle columns of this table show the comparison between the use of these conjunctions in the 10 folios based on

TABLE V

Frequency of Occurrence of Conjunctions in Galileo's Early Latin MSS
(Number of occurrences per thousand words)

PREFERRED CONJUNCTIONS	MS Gal 27*	MS Gal 46† (16v–26r)	MS Gal 46† (entire)	MS Gal 71†
nam	2.3	6.4	4.1	2.4
et	24.8	31.5	37.9	24.9
ut	9.2	12.0	11.6	12.4
cum	3.8	6.9	6.2	7.7
sed	6.6	3.9	5.9	6.3
neque	3.2	1.8	2.3	1.6
quia	16.9	3.9	11.8	4.9
NON-PREFERRED CONJUNCTIONS				
enim	1.1	1.2	1.8	7.7
quare	1.1	0.0	0.8	1.8
quoniam	0.0	0.0	0.0	0.0
deinde	0.3	0.9	0.6	1.3
denique	0.1	0.5	0.2	0.0
unde	0.4	0.2	1.1	0.6

* Word counts are based on a transcription of the text by Adriano Carugo.
† Word counts are based on Favaro's reading in *Opere* 1.

Clavius and in the entire composition, and the agreement here is very good. Indeed, the only serious departure is in the less frequent use of *quia*, for which Galileo seems to have a more decided preference than is obvious from his conversion index for this word based on Clavius. Clavius, in fact, prefers more sophisticated modes of expression, and so does not use the simple *quia* when giving reasons – a factor that apparently inhibited Galileo's use of it in this portion of his composition.

When the analysis is extended to Galileo's other Latin manuscripts dating from his Pisan period, moreover, additional confirmation for a characteristic style is given. These results are presented in the two outer columns of Table V. MS Gal 27 is a set of notes in Galileo's hand not unlike the two sets contained in MS Gal 46, except that it is concerned with matters relating to Aristotle's *Posterior Analytics*; MS Gal 71, on the other hand, includes all of Galileo's early writings on motion, his rather extensive *De motu antiquiora*, commonly dated by historians between 1590 and 1592. The latter work, in particular, has always been regarded as Galileo's own composition, and since essentially the same preferences are regulative throughout all of these works, it seems on face value that Galileo was the author who put them all together.

These observations, based as they are on word counts, at best have suasive force, but they provide some basis for maintaining that all three Pisan manuscripts are the work of Galileo's head as well as his hand, and that they probably were composed by him, as will be argued in more detail below, in connection with his Pisan professorship from 1589 to 1591.

To return, then, to the portion of MS Gal 46 based on Clavius, other *prima facie* evidences can be adduced to support the view that Galileo himself actually used Clavius's text when composing these notes, and that they are not therefore the work of some intermediate author merely copied by Galileo as a scholastic exercise. Clavius's commentary went through five editions, the first appearing in 1570, but there are indications that Galileo made use of the second edition of 1581 or possibly a later printing.[20] A likely account of how Galileo worked is suggested by Plate I, which shows folios 17r and 18r of Galileo's manuscript juxtaposed with pp. 62, 63, and 46 of the 1581 edition. On this illustration passages have been identified that show errors in copying, where, for example, Galileo copied from the wrong line, crossed out his mistake, and then wrote the passage correctly, or changed his copy to agree with the text. While it is possible that Galileo could have done this from a

PLATE I

MS Gal 46 fol. 17r:

Clavius, In sphaeram... (1581)

p. 62:

PROPTER QVAE PHAENOMENA ASTRONOMI *motum trepidationis stellis fixis attribuerint.*

QVONIAM vero supra dictum est, stellas fixas non solum duplici isto

p. 63: — copied wrong place, crossed out

motu, quorũ vnus est ab ortu in occasum, alter vero ab occasu in ortum, moueri, sed habere etiam propriũ motum accessus & recessus, quem trepidationis dicunt: ostendendum nunc est, quæ phænomena, apparentiæve Astronomos coegerint, vt hunc motum in cœlo ponerent: Non pauci enim motum hunc omnino explodendum à scholis Astronomorum, tanquam ridiculum, arbitrãtur. Primo ergo obseruarũt, stellas fixas inæqualiter incedere ab occidẽte in orientem: Nunc enim velocius, nunc tardius, nunc nullo pacto moueri in Zodiaco videbantur, nunc vero retrocedere ab oriente in occidentem, præter illum motum diurnum, & eandem nihilominus distantiam a centro mundi habere. Quare dixerunt eas moueri à septentrione in austrum & contra, vt supra declaratum fuit in motu illo accessus & recessus. Propter hunc enim mo-

— errors in copying, corrected to agree with text

fol. 18r: — wrote "simiae" incorrectly

p. 46:

gradi equi velocissimi procreãtur, & validissimi, qui in alijs regionibus eiusdem latitudinis minime producũtur. Denique in Mauritania innumeræ quasi simiæ generantur: Et multa alia huiusmodi experimenta ad luci possent, vt à vitibus, arboribus, fructibus, &c. qui omnes varij effectus à cælo duntaxat quiescente produci videntur. Scio philosophos respondere, hãc diuersitatem effectuum in eodem climate pedere totam ex varia dispositione terræ: sed instant auctores prædicti, cum terra disponatur varie à varijs aspectibus corporum superiorum, non poterit reddi sufficiens causa, cur in eodem climate eadem non sit dispositio, quandoquidem omnes partes eiusdem climatis respectu cælorum mobilium eosdem habeant aspectus successiue. Verum enim vero quidquid dicatur hac de re, hoc certum esse debet, sine magna temeritate

— copied from line above, then corrected

— copied incorrectly, crossed out and repeated

manuscript source rather than from the printed text here illustrated, on face value one would be disposed to see Clavius's commentary as his direct source.[21]

To come now to the dependence on Pererius, suspected by myself and independently identified by Carugo, this is somewhat similar to that on Clavius but it is not so striking – the passages used are neither as extensive nor as concentrated as those used for the 10 folios. At one place in MS Gal 46, however, Galileo makes explicit reference to Pererius's text; this is at the end of the first question in the *Tractatio de mundo*, which discusses the opinions of ancient philosophers concerning the universe. There Galileo concludes his exposition with the words:

The opinion of Aristotle is opposed to the truth – for his arguments and those of Proclus, Averroës, and others supporting the eternity of the world, together with the answers, read Pererius, Book 15.[22]

Checking this citation, one finds in Book 15 of *De communibus* a considerable discussion paralleling closely the material found in the fourth question of Galileo's *Tractatio de mundo*, entitled 'Whether the world could have existed from eternity?' This occurs in Chapter 10 of that book, also entitled 'Whether the world could have existed from eternity,' which begins with a 'first opinion' that is quite similar to the 'third opinion' Galileo gives in answer to the same query.[23] Below are reproduced in parallel column excerpts from the two Latin texts, to show the resemblance between them:

PERERIUS	GALILEO
De communibus. . .principiis	MS Gal 46
Lib. 15. De motus et mundi aeternitate	Tractatio prima de mundo
Caput decimum	Quaestio quarta
An mundus potuerit esse ab aeterno?	An mundus potuerit esse ab aeterno?
Prima opinio est Henrici de Gandavo, Quod. 1, quaest. 7, et Philoponi in eo libro quo respondet argumentis Procli pro aeternitate mundi. . .	Tertia opinio est Philoponi in libro quo respondet argumentis Procli pro aeternitate mundi, Gandavensis in quodlibeto p°, . . .
Quod probant his rationibus, 1. creatio	Probatur haec opinio. . .2° quia creatio

est productio ex nihilo. . .necesse est in creatione non ens et nihil praecedere ipsum esse rei quae creatur. . .	est productio ex nihilo: quo fit ut non ens necessario debeat praecedere rem creatam. . .
2. De omni producto verum est dicere ipsum produci, . . .sed de creatura non est verum dicere ipsam semper produci; alioquin nulla creatura habere esse permanens, sed tantum esse successivum. . . ergo si in aliquo instanti producitur, non semper, nec ab aeterno habet esse.	Confirmatur: quia de eo quod producitur verum est dicere producitur: sed creatura quae producitur non potest semper produci, alioqui [sic] non haberet esse suum permanens sed successivum. . .ergo in instanti producitur, et, ut est consequens, non potest ab aeterno esse.
3. Creatura habet esse acquisitum a Deo, ergo habet esse post non esse, nam quod acquirit aliquid, id non semper habuit. . . .	Confirmatur adhuc: quia creatura habet esse acquisitum a Deo; ergo habet illud post non esse; quia non acquiritur quod habetur, sed quod non habetur. . . .

Here a careful study of the Latin texts shows a close dependence of MS Gal 46 on Pererius's text completely analogous to the dependence noted earlier on Clavius. The arguments have been rearranged, some of the words have been changed and the inflections have been altered, but the content is essentially the same. Thus the conclusions that have been reached above with regard to Clavius's text would seem to apply *a pari* to Pererius's. The works of the two Jesuits explicitly cited in Galileo's notes, the two most recent books in all the works he there references, would therefore appear to have supplied a significant portion of the information they contain.

IV. MS GAL 46: MANUSCRIPT SOURCES

The foregoing researches are of paramount importance; not only do they throw unexpected light on possible sources of Galileo's early notebooks, but they register a considerable advance over Favaro's conjecture that the notes were based on the teaching of Buonamici – whose writings contains nothing comparable to the materials just discussed. On one matter, however, they are not particularly helpful, and this is Galileo's references to the *Doctores Parisienses*. Now it is significant that Galileo mentions the *Parisienses* in only two places in MS Gal 46, and that one of these citations occurs in precisely the question just discussed that seems to be based on Pererius. A careful search of Pererius's text, on the other hand, shows no mention of the *Doctores*

Parisienses in this or in any other locus. Where, therefore, did Galileo get this citation? Or did he in fact use Pererius as a guide and actually return to primary sources himself, as Duhem claimed in the first place? Allied to these questions are others relating to the composition of MS Gal 46 as a whole. Already noted is the fact that the printed works of Clavius and Pererius can account for 15% of the contents of these notes. That leaves 85% of the notes unaccounted for. What was *their* source? Or could Galileo possibly have composed them himself?

Fortunately, to answer this last question there is available one of Galileo's Latin compositions known beyond all doubt to be original, the autograph of the famous *Nuncius sidereus* written by him in 1609. Although the subject matter of this work was quite familiar to Galileo, its actual writing in Latin proved somewhat tortuous: as can be seen on Plate II, which duplicates the first page of his composition, almost every line has cross-outs and corrections, there are frequent marginal inserts, and the manuscript as a whole is decidedly messy. When this Paduan composition is compared with the Pisan manuscripts, in fact, the latter are found to be quite neat and clean. They have their evidences of copying, as already remarked, and on most pages there are a few words crossed out, others inserted, frequent superscript or subscript modifications of inflection, minor deletions, and so forth, but nothing of the magnitude of the marking-up found in Galileo's original Latin composition. Favaro's paleographical comments thus appear to be correct: it seems probable that *all* of MS Gal 46 is derivative, although not in the precise way Favaro thought. To verify this a careful check of all 97 folios was made, noting the number of corrections on each page and the evidences of copying they reveal, to see if these were concentrated in places for which the source was known or were distributed uniformly throughout the entire composition. Again the results, summarized on Plate III, are surprising. The 10 folios based on Clavius and the others based on Pererius are not markedly different from any other folios; indeed those toward the end of the manuscript, not yet discussed, give even more indications of copying than do the earlier ones. Moreover, some of the cross-outs and changes in this part of the manuscript raise the question whether Galileo had difficulty not merely with the subject matter, but even with reading the text from which he was working. This query further suggests unfamiliar handwriting: perhaps Galileo based these notes not on printed sources alone but on partially illegible manuscript sources as well?

SIDEREUS NUNCIUS -- AUTOGRAPH OF FIRST PAGE:

PLATE II

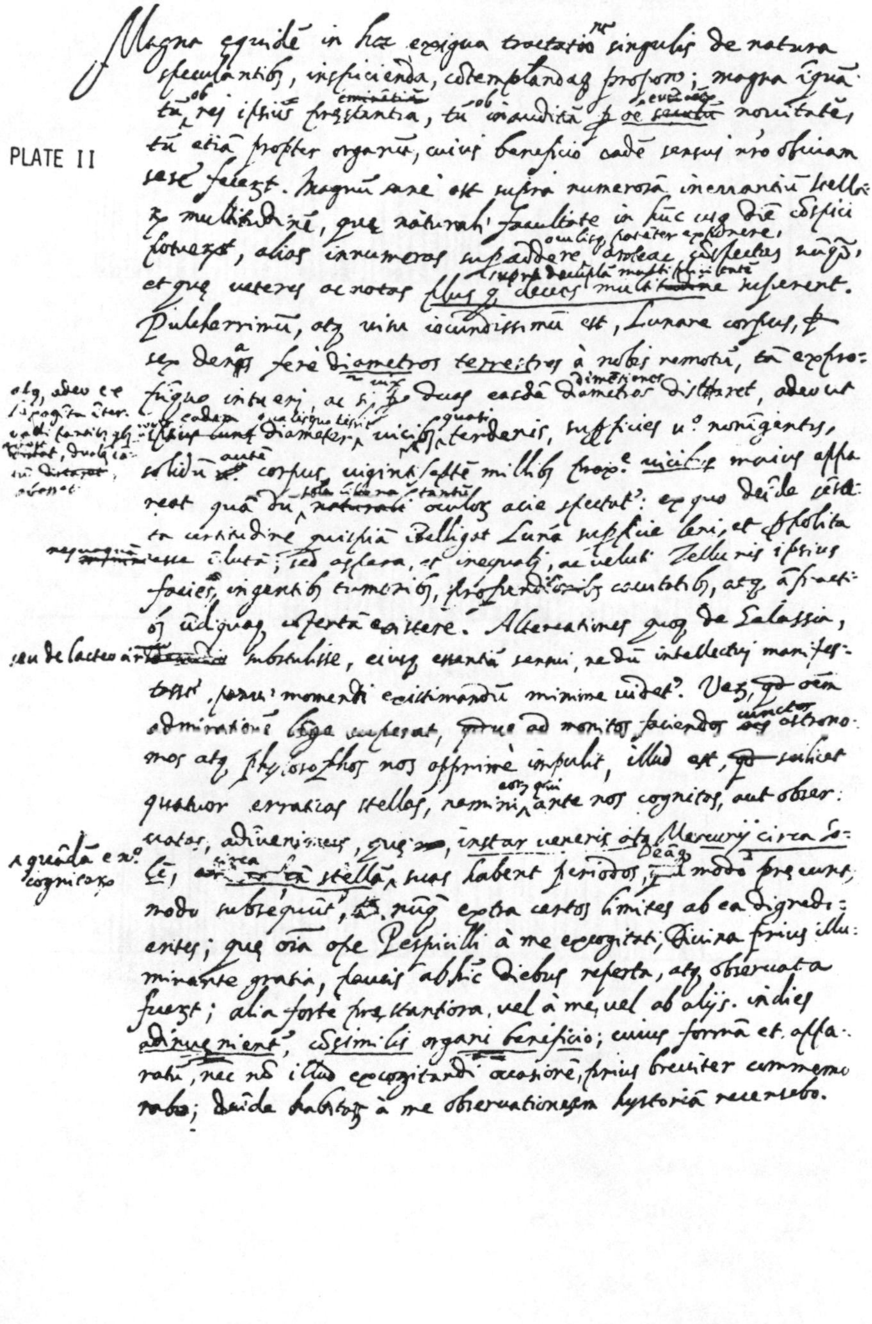

Magna equidē in hac exigua tractatiōne singulis de natura
speculantibus, inspicienda, cōtemplandaq propono; magna inquā
tū ob rei ipsius præstantiam, tū ob inauditā per ōe ævum novitatē,
tū etiā propter organū, cuius beneficio eadē sensui nrō obviam
sese fecerūt. Magnū sane est supra numerosā inerrantiū stella-
rū multitudinē, quæ naturali facultate in hūc usq diē cōspici
poterant, alias innumeras superaddere, oculisq palam exponere, [illegible]
et quæ veteres ac notas plus q decies multitudine superent.
Pulcherrimū, atq visu iocundissimū est, Lunare corpus, per
sex denas ferè diametros terrestres à nobis remotū, tā ex pro-
pinquo intueri, ac si per duas easdē dimensiones distaret, adeo ut
[illegible] diameter vicies, superficies vero noningentis,
solidū autem corpus, vigintiseptem millibus fere vicibus maius appa-
reat quā dū solo libero oculo aspectatur: ex quo deinde certa
cū certitudine quispiam intelligat Lunā superficie levi, et perpolita
[illegible] esse indutā; sed asperā, et inæqualē, ac veluti Telluris ipsius
facies, ingentibus tumoribus, profundis lacunis, atq anfracti-
bus undiquaq confertā existere. Alterationes quoq de Galaxia,
seu de lacteo [illegible] sustulisse, eiusq essentiā sensui, nedū intellectui manifes-
tasse, parui momenti existimandū minime videtur. Verū, quod oēm
admirationē longe superat, quodq ad monitos faciendos cunctos astronomos atq philosophos nos apprime impulit, illud est, quod scilicet
quatuor erraticas stellas, nemini ante nos cognitas, aut obser-
vatas, adinvenimus, quæ, instar Veneris atq Mercurij circa So-
lē, circa [illegible] stellā suos habent periodos, [illegible] modo præcurrentes,
modo subsequentes, nunquā extra certos limites ab ea digredi-
entes; quæ oīa ope Perspicilli à me excogitati, divina prius illu-
minante gratia, paucis abhinc diebus reperta, atq observata
fuerūt; alia forte præstantiora, vel à me, vel ab alijs. in dies
adinvenientur, cōsimili organi beneficio; cuius formā et appa-
ratū, nec nō illud excogitandi occasionē, prius breviter commemo-
rabo; deinde habitarū à me observationū historiā recensebo.

[Margin:] atq, adeo ut e [illegible]

[Margin:] a quādā e nō cognitas

PLATE III

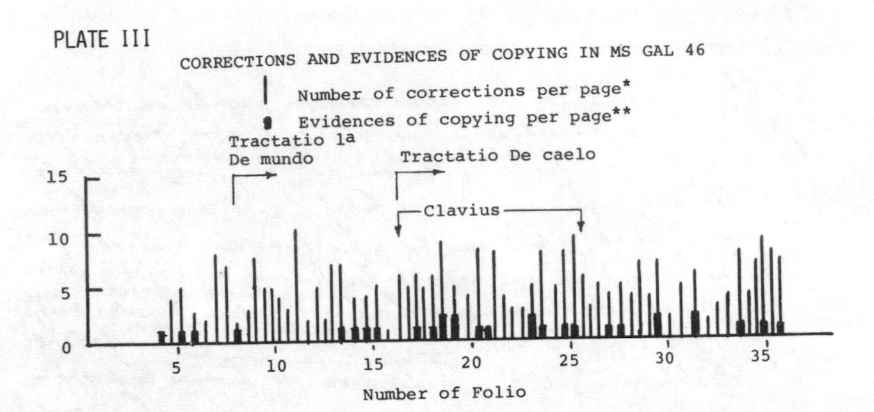

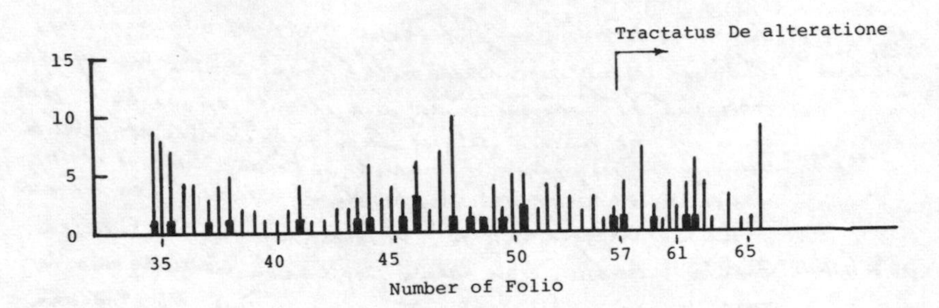

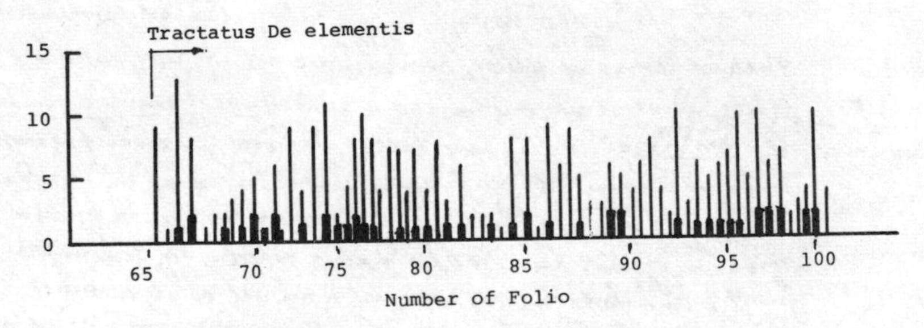

*Total = 945 **Total = 88

Further investigation, following up this line of reasoning, showed that Pererius had taught the entire course of natural philosophy at the Collegio Romano in 1565–1566, and that handwritten copies or reports (*reportationes*) of parts of his course during that academic year are still extant. Not only this, but a considerable collection of similar lecture notes from the Collegio have been preserved, spanning the interval from the 1560's through most of the seventeenth century. It turns out that the Jesuit professors who taught natural philosophy at the Collegio generally did so for only one year, although occasionally they repeated the course; in most instances, moreover, they had taught the entire logic course the year previous, ending with the *Posterior Analytics.*[24] Many of those who taught these courses prepared questionaries on the relevant works of Aristotle; professors of natural philosophy did so for the *Physics*, the *De caelo*, and the *De generatione*, and in some cases they even reached the *Mateorology* – precisely the span of subject matter envisaged in Galileo's enterprise. Unfortunately not all of these questionaries have been preserved; thus far the writer has located only the following from the period of Galileo's lifetime:

Year	Professor	Notes
1565–1566	Benedictus Pererius	*Physics, De caelo, De generatione*
1567–1568	Hieronymus de Gregorio	*Physics, De caelo, De generatione*
1588–1589	Paulus Valla	*Meteorology*
1589–1590	Mutius Vitelleschi	*De caelo, De generatione, Meteorology*
1597–1598	Andreas Eudaemon-Ioannes	*Physics, De caelo, De generatione*
1603–1604	Terentius Alciati	*De cometis*
1623–1624	Fabius Ambrosius Spinola	*Physics, De generatione*
1629–1630	Antonius Casiglio	*Physics, De caelo, De generatione*

Further research will undoubtedly turn up other Collegio *reportationes* for courses taught in the years between 1560 and 1642, for they are known to be diffused widely through European manuscript collections. For the time being, however, it should be noted that all of the first four authors on the list above have teachings that can be found in MS Gal 46, and that particularly the notes of the third and fourth, i.e., Paulus Valla and Mutius Vitelleschi, contain substantial amounts of Galileo's material. Indeed, when these manuscript sources are taken in conjunction with the printed sources already discussed, practically the entire contents of MS Gal 46 can be seen to derive, in one way or another, from courses taught at the Collegio Romano up to the year 1590. This, then, supplies the clue to Galileo's otherwise odd citation of sources as given in Table I: trained both scholastically and humanistically, the Jesuit professors from whose writings Galileo worked were well acquainted with medieval sources, and particularly with Thomas Aquinas and 'the Thomists,' but they also knew the Averroist thought then being propagated in the universities of Northern Italy, and of course they were no strangers to the primary sources of classical antiquity.

To give some idea of these manuscript sources, there are reproduced below transcriptions of portions of two of the *reportationes* that could well be the loci from which Galileo obtained his references to the *Doctores Parisienses*. The first is that of Vitelleschi, who lectured on the *De caelo* during the academic year 1589–1590. The question wherein the citation occurs is the same as that already seen in Galileo and Pererius, namely, whether the world could have existed from eternity. Galileo gives four different opinions on the question, whereas Vitelleschi gives only three; a close examination of the first of Vitelleschi's opinions, however, shows that it can be divided into two interpretations, and it is a simple matter to make a fourth opinion out of one of the alternatives. Portions of the Latin text of all four of Galileo's opinions are given below on the right, and in parallel column on the left are shown excerpts from Vitelleschi's notes, rearranged where necessary to show similarities in the text:

VITELLESCHI	GALILEO
APUG/FC 392 (no foliation)[25]	BNF MS Gal 46, fol. 13r–15r
Tractatio prima. De mundo.	Tractatio prima. De mundo.
Disputatio quarta.	Quaestio quarta.
An mundus potuerit esse ab aeterno?	An mundus potuerit esse ab aeterno?
Prima sententia { ita Gregorius in 2° dist. p^{a} quest. 3^{a} art. p^{o} et 2°, Occam ibidem. . .Gabriel. . .hi tamen duo defendunt hanc sententiam tamquam probabiliorem, et volunt contrariam etiam probabiliter posse defendi; idem sentiunt Burlaeus et Venetus in t. 15 8^{i} Phys.,	Prima opinio est Gregorii Ariminensis, in 2° dist. p^{a} q^{e} 3 art. p^{o} et 2°, Gabrielis et Occam ibidem; quamvis Occam non ita mordicus tuetur hanc sententiam quin etiam asserat contrariam esse probabilem. Hos secuntur Ferrariensis, in 8 Phys. qu. 15, et Ioannes Canonicus, in

Canonicus ibidem. . .Eandem sententiam sequuntur Ferrariensis 8 Phys. quest. 3^{a}, . . .Soncinas. . .Herveus. . .Capreolus}[26] vult mundum. . .secundum omnia entia que continet, tam permanentia quam successiva, tam corruptibilia quam incorruptibilia, potuisset esse ab aeterno.	8 Phys. q. p^{a}, et plerique recentiorum, existimantium mundum potuisse esse ab aeterno, tam secundum entia successiva quam secundum entia permanentia, tam secundum corruptibilia quam secundum incorruptibilia.
Secunda sententia {ita Durandus in 2° dist. p^{a} quest. 2^{a} et 3^{a}, quem sequuntur plurimi recentiores; Sotus vero 8 phys. quest. 2^{a}. . .} concedit res incorruptibiles potuisse esse ab aeterno, negat tamen res corruptibiles. . .	Secunda opinio est Durandi in 2° dist. p^{a}, quem secuntur permulti recentiorum; et videtur etiam esse D. Thomae, in p^{a} parte q. 46 art. 2°, sentientis aeternitatem repugnare quidem corruptibilibus, non tamen incorruptibilibus. . .
Tertia sententia {ita Philoponus contra Proclum sepe. . .Enricus quodlibeto p^{o} . . .Marsilius in 2° dist. p^{a} quest. 2^{a} art. p^{o}, S. Bonaventura ibidem parte p^{a} art. p^{o} quest. 2^{a}. . .plurimi S. Patres quos infra citabo} . . .	Tertia opinio est Philoponi in libro quo respondet argumentis Procli pro aeternitate mundi, Gandavensis in quodlibeto p^{o}, D. Bonaventura in 2° dist. p^{a} quo^{e} 2^{a}, Marsilii in 2° dist. p^{a} art. 2°, Burlei in 8 Phys. in quo^{e} hac de re super t. 15; et Sanctorum Patrum. . .
[Prima sententia:] Cum hac sententia convenit S. Thomas ut colligitur ex opusculo de hac re et. . .alibi {Scotus autem in 2° dist. p^{a} quest. 3^{a} neutram partem determinate sequitur sed putat utramque esse probabilem. Quod etiam *docent Parisienses 8 Phys. quest.* p^{a}. . .}	Quarta opinio est D. Thomae, in q^{e} 3^{a} De potentia art. 14, et in opusculo De aeternitate mundi, et in 2° Contra gentiles c. 38, et in p^{a} parte q. 46, Scoti in 2° dist. p^{a} q. 3, Occam in quodlibeto 2° art. 5, *doctorum Parisiensium 8 Phys. q.* p^{a}, Pererii in suo 15, et aliorum. . .

What is most striking about these parallels, of course, is that practically every authority and locus cited by Vitelleschi is also in Galileo, usually in abbreviated form. And, most important of all, the *Parisienses* are mentioned explicitly by Vitelleschi, and in a completely similar context, although for Galileo this occurs in his fourth opinion whereas for Vitelleschi it is in a variation of his first. More than that, although Galileo cites Pererius immediately after the *Parisienses*, Galileo's enumeration of opinions does not follow Pererius's enumeration but rather Vitelleschi's.[27] In light of this discovery one may now question whether Galileo did copy from Pererius after all, or whether his proximate source was a set of notes such as Vitelleschi's, which in turn could have taken into account the exposition of Pererius.[28]

Galileo's other mention of the *Doctores Parisienses* occurs in the second set of notes that make up MS Gal 46, more specifically in his Treatise on the Elements, where he makes the brief statement:

We inquire, second, concerning the size and shape of the elements. Aristotle in the third *De caelo*, 47, and in the first *Meteores*, first summary, third chapter, followed by the *Doctores Parisienses*, establishes a tenfold ratio in the size of the elements, i.e., water is ten times larger than air, and so on for each. Actually, however, whether this is understood of the extensive magnitude of their mass or of the magnitude and portion of their matter, I will show elsewhere by mathematical demonstration that it is false. . . .[29]

Now Vitelleschi in his lecture notes on *De generatione* mentions the problem of the size and shape of the elements, but says that he treats it elsewhere,[30] and so he is of no help on this particular citation. But Paulus Valla, who taught the course in natural philosophy the year before Vitelleschi, i.e., in 1588–1589, wrote an extensive commentary on Aristotle's *Meteorology* to which he appended a Treatise on the Elements, and in the latter there is a passage that is similar to Galileo's. Again the two are reproduced below in parallel column, with Valla's text on the left and Galileo's on the right:

VILLA	GALILEO
APUG/FC 1710 (no foliation)	BNF/MS Gal 46, fol. 76r
Tractatus de elementis	Tractatus de elementis
De quantitate elementorum	De magnitudine et figura elementorum
Aliqui existimant elementa habere inter se decuplam proportionem, ita ut aqua sit decuplo maior terra, aer decuplo maior aqua, [etc.]. . .ita tenent *Doctores Parisienses*, p^{o} Meteororum, q. 3, . . .Probatur ex Aristotele 3° Caeli, t. 47, . . .p^{o} Meteororum, summa p^{a}, cap. 3. . .Contrariam sententiam habent communiter omnes mathematici et multi etiam Peripatetici non dari scilicet in elementis ullam determinatam proportionem in quantitate.	Aristoteles in 3° Caeli, 47, et p^{o} Meteororum, summa p^{a}, cap^{e} 3°, quem secuti sunt *Doctores Parisienses*, in magnitudine elementorum constituit proportionem decuplam: idest, aqua sit decies maior quam terra; et sic de singulis. Verum, hoc, sive intelligatur de magnitudine molis extensiva, vel de magnitudine et portione materiae illorum, demonstrationibus mathematicis alibi ostendam id esse falsam.

Note here exactly the same reference to the *Doctores Parisienses*, as well as the same peculiar way of citing Aristotle's *Meteorology*. Also interesting is Galileo's statement that he is going to disprove the *Parisienses*' opinion elsewhere 'by mathematical demonstration,' which in fact he does not do, at least not in the portions of his notes that have come down to us; Valla, on the other hand, gives all the arguments of the 'mathematicians' that go counter to the *Parisienses*, and thus could have been a source for Galileo's projected demonstrations.[31] And needless to say, if Galileo had access to Valla's notes or to others like them, he could easily have written what he did without ever having looked at these authors from fourteenth-century Paris.[32]

Before leaving the subject of manuscript sources for MS Gal 46, it may be well to raise the question of possible intermediates for the portions of the notes apparently based on Clavius. Galileo's presentation differs most markedly from Clavius's in the simple fact that it locates many details of Ptolemaic astronomy, not in an exposition of the *Sphere* of Sacrobosco, as does Clavius, but rather in notes on Aristotle's *De caelo*. Less striking is an odd interpolation apparently introduced by Galileo in this material and already mentioned above, i.e., his citation of St. John Chrysostom in the first opinion on the number of the heavens. These differences suggest two apparently innocuous questions: (1) Did Galileo have any precedent for treating the number and ordering of the heavenly orbs in questions on the *De caelo*; and (2) Where did Galileo obtain his reference to Chrysostom, a Church Father about whom he would not be expected to know much in his own right, when such a reference is not found in any edition of Clavius? Now suggestions for answers to these may be found in the questionaries of Hieronymus de Gregorio, who taught the course in natural philosophy at the Collegio in 1568. Among his questions on the second book of the *De caelo* are two entitled respectively *De numero orbium caelestium* and *De ordine orbium caelestium*. Note here that the first title is exactly the same as that in Clavius, whereas the second differs by only one word: Clavius's title reads *De ordine sphaerarum caelestium*, i.e., it substitutes *sphaerarum* for *orbium*. Returning now to Galileo's notes, we find that his second question in the *Tractatio de caelo* follows De Gregorio's reading, not Clavius's. And more remarkable still, among the many opinions mentioned by De Gregorio at the outset of his treatment of the first topic is to be found the name of John Chrysostom. De Gregorio's notes, of course, are very early, 1568, composed apparently before the first edition of Clavius,[33] and one need not maintain that Galileo actually used these *reportationes* when composing his own notes. But it could well be that a tradition of treating some astronomy in the course *De caelo* already existed at the Collegio Romano, and that another Jesuit (yet unknown) had previously culled from

Clavius's *Sphaera* the type of material contained in MS Gal 46, thus serving as the proximate source of Galileo's note-taking.[34] This still leaves open, to be sure, the question of multiple sources for MS Gal 46, and it also allows room for Galileo's distinctive style of Latin composition to show through the final result, as explained above.[35]

V. MS GAL 46: THE PROBLEM OF DATING

All of this evidence makes highly plausible one aspect of Favaro's critique of Duhem, i.e., that MS Gal 46 derives from secondary rather than from primary sources. In another respect, however, Favaro might have erred, and this in dating the manuscript's composition in 1584, particularly considering the late date of the Valla and Vitelleschi materials, written as these were in 1589 and 1590 respectively. More pointedly, perhaps the earlier curator was closer to the truth when indicating that the notes were composed 'around 1590.' If so, in dating them too early Favaro could well have underestimated their role in Galileo's intellectual development, and so overlooked an important element of truth in Duhem's continuity thesis. On this score it is important to turn now to the problems of dating, to the evidence Favaro considered indisputable for the year 1584, and to a possible reinterpretation of that evidence.

First it should be remarked that it is practically impossible to date MS Gal 46 as a juvenile work on the basis of Galileo's handwriting alone. To the writer's knowledge only five Latin autographs of Galileo are still extant in various collections, and portions of each of these are duplicated on Plate IV. At the top are the opening lines of the draft of *Sidereus nuncius* written by Galileo – and this is the only date known for certain – in 1609, at the age of 45. Just under this is an extract from the notes on motion, the *De motu antiquiora* of MS Gal 71, usually dated between 1590 and 1592. Under this again are a few lines from MS Gal 46, whose writing was put by Favaro in 1584. Under this yet again is an extract from the smaller set of notes in MS Gal 27 relating to Aristotle's *Posterior Analytics*, which seem to be earlier than those in MS Gal 46; because of spelling errors they contain Favaro regarded them as Galileo's first attempt at Latin scholastic composition.[36] The bottom sample, finally, contains a few lines from a brief exercise of nine folios (at the back of MS Gal 71), wherein Galileo is translating from Greek to Latin a passage attributed to the Greek rhetorician Isocrates. These

PLATE IV GALILEO'S LATIN HANDWRITING 1579?-1609

1609
MS 48
fol.
8r

Magna equidē in hac exigua tractatione singulis de natura speculantibus, inspicienda, cōtemplandaque propono; magna inquā tū ob rei ipsius praestantia, tū ob inaudita per aevum novitate, tū etiā propter organū, cuius beneficio eadē sensui nro obviam

MS 71
fol.
3v

Motus violentus à quo fiat
Utrum medium sit necessarium ad motū
an detur simpliciter grave et simpliciter
an elementa in proprio loco sint gravia aut levia

gravis in aliquo [illegible] bilis [illegible]

MS 46
fol.
33r

quarto
Coroll: primum

[illegible]. 4. ex etimologia nominis dicitur enim ether ab eo quod semper currat. et ex his colligimus celum esse inalterabile tum quia alteratio requirit contrarium nam alteratio non fit sine contrariis aliqua qualitas opposita, tum quia in celo non reperiuntur qualitates alterativae, tum demum quia alteratio est dispositio ad generationem et corruptionem.

MS 27
fol.
4r

1a Con. 1a principia sunt aliquo modo cognoscenda ad [illegible] communiter [illegible] ex demonstratione [illegible] principia [illegible] regulariter illius [illegible] aliquo modo pendet [illegible] aliquo modo a primis principiis [illegible]

Concl.o

1579?
MS 71
fols.
132r
132v

[illegible] ingenia [illegible]
[illegible] πολὺ δὲ μεγίστην διαφορὰν εἰλήφασιν ἐν τοῖς
quapropter nos non adortationem invenientes, sed admonitionem conscribentes
διόπερ ἡμεῖς οὐ παράκλησιν εὑρόντες, ἀλλὰ παραίνεσιν γράψαντες

particular folios show evidence of an older hand correcting Galileo's translation, a fairly good indication that they are a student composition.[37] Presumably Galileo wrote this exercise during his humanistic studies, before he entered the University of Pisa in 1581; it could date from as early as 1579, when he was only 15 years of age.

In all of these samples of Galileo's Latin hand, and they span a period of almost 30 years, certain similarities are recognizable. For example, his capital 'M' remains pretty much the same; the way he links the 'd-e' and the 'd-i' is constant; the abbreviation for 'p-e-r' does not change; his manner of writing 'a-d' varies only slightly. Galileo's hand is consistent and in no way idiosyncratic; indeed, it resembles many other Tuscan hands of the late *cinquecento*. And, from a comparative point of view, the hand that wrote MS Gal 46 is no more patently juvenile than the one that penned the opening lines of the *Sidereus nuncius* or the extract from the *De motu antiquiora*. Thus it is not transparently clear that these notes can be written off as a youthful exercise, at least on the basis of the hand that wrote them.

To come now to Favaro's internal evidence, in the second question of the *Tractatio de mundo* Galileo considers the age of the universe. This particular question arose because of Aristotle's teaching that the universe is actually eternal – not the same problem as that already mentioned, whether it *could be* eternal, which questions the possibility rather than the actuality of its eternal existence. Galileo's answer is orthodox, in conformity with the Church's teaching on creation: Aristotle is wrong, the universe was created in time. He goes on:

> To anyone inquiring how much time has elapsed from the beginning of the universe, I reply, though Sixtus of Siena in his *Biblioteca* enumerates various calculations of the years from the world's beginning, the figure we give is most probable and accepted by almost all educated men. The world was created five thousand seven hundred and forty-eight years ago, as is gathered from Holy Scripture. For, from Adam to the Flood, one thousand six hundred and fifty-six years intervened; from the Flood to the birth of Abraham, 322; from the birth of Abraham to the Exodus of the Jews from Egypt, 505; from the Exodus of the Jews from Egypt to the building of the Temple of Solomon, 621; from the building of the Temple of Solomon to the captivity of Sedechia, 430; from the captivity to its dissolution by Cyrus, 70; from Cyrus, who began to reign in the 54th Olympiad, to the birth of Christ, who was born in the 191st Olympiad, 560; the years from the birth of Christ to the destruction of Jerusalem, 74; from then up to the present time, 1510.[38]

Now that is a fascinating chronology, and one can well imagine Favaro's excitement when he transcribed this passage and saw in it a definitive way of dating the composition of these notes. All we need do, said he, is focus on the last two figures Galileo has given, for these, when added together, give the interval from the birth of Christ to Galileo's writing. So add 74 to 1510, and we obtain the result desired, *Anno Domini* 1584. That is the year in which the notes were obviously composed.[39]

This, of course, is a piece of evidence to be reckoned with. But Favaro's interpretation of it is not the only one possible. To permit a different reckoning the crucial point to notice is that the foregoing chronology does not supply a single absolute date; all that it records are intervals between events. Now, if one can accurately date any event in Galileo's account, then of course he can calculate other dates from these intervals. For example, if one knows the year in which the destruction of Jerusalem took place, then by adding 1510 to that date he will determine the year referred to in the notes as 'the present time.' As it just happens, the best known date of all the events mentioned, and one confirmed in secular history, is the date of the destruction of Jerusalem, which took place in A.D. 70.[40] So, add 1510 to 70 and the result is 1580, not 1584, as the date referred to as 'the present time.' Moreover, if Galileo wrote these notes in 1580, he was then a mere sixteen years of age, not yet a student at the University of Pisa, and Favaro's title would become even more appropriate than he thought – they would be *juvenilia* beyond all question and doubt.

What this observation should serve to highlight is a more problematic dating that lies behind Favaro's calculation, namely, determining the year in which Christ was born. Favaro simply assumed that Christ's birth was in A.D. 0, but scholars who were writing chronologies in the 1580's would permit themselves no such assumption. Most of their attention was devoted to calculating the years that transpired between the creation of the world and the birth of Christ so as to establish the time of Christ's birth in relation to the age of the universe. Going back, then, to Galileo's epochs and summing the intervals he gives from Adam to the birth of Christ, one obtains a total of 4164 years. Now, it is an interesting fact that throughout the course of history hundreds of calculations of this particular sum have been made, and despite Galileo's saying that his figures are "most probable and accepted by almost all educated men," no recent historian has uncovered any chronology that gives Galileo's implicit sum, 4164 years from the world's creation to the birth of Christ. The chronologer William Hales, in his *New Analysis of Chronology*, printed in 1830, gives more than 120 different results for this computation, and says that his list "might be swelled to 300" without any difficulty.[41] Some of his figures will be of interest to historians of science: they range from the calculation of Alphonsus King of Castile, in 1252, who gives 6984 as the total number of years between creation and Christ's birth, to that of Rabbi Gerson, whose sum is only 3754. The calculators include some eminent figures in science's history: Maestlin, who gives 4079 years; Riccioli, who computes 4062; Rheinhold, who has 4020; Kepler, who gives 3993; and Newton, who computes only 3988.[42] Such calculations, needless to say, are very complex and require a detailed knowledge of Scripture to be carried out. It is quite unlikely that Galileo would have been able to do this himself; more probably he copied his result from another source. But again, the identity of that source has been a persistent enigma, for no chronology in Hales's list or elsewhere yields Galileo's precise result.[43]

Here again a possible answer can be found by employing the procedure that worked with the *Doctores Parisienses*, i.e., by turning to the Jesuits of the Collegio Romano. We have already discussed Benedictus Pererius, who taught the course in physics at the Collegio between 1558 and 1566, at which time he disappeared from the *rotulus* of philosophy professors and began teaching theology.[44] In 1576, moreover, he was the professor of Scripture at the Collegio, a post he held until 1590, and again in 1596–97.[45] Only fragments of his Scripture notes survive, but fortunately for our purposes he published two Scriptural commentaries, one on the Book of Daniel printed at Rome in 1587, the other on the Book of Genesis printed there in 1589.[46] When these are studied for evidences of a chronology of creation it is found that the chronology recorded by Galileo, with the exception of only a single interval, employs the figures given by Pererius. Thus this computation, like practically all else in MS Gal 46, derives from work done at the Collegio Romano and, in this case, recorded in a printed source not available until the year 1589.

The details of these calculations are given in Table VI and its accompanying notes, but a few additional observations may help locate these materials in context. The problem with intervals of the type shown on Table VI is that not all Scripture scholars choose precisely the same events for their computations; depending on their knowledge of particular books of the Bible they find some intervals easier to compute than others. But a more or less standard chronology, standard in the sense that it was widely diffused and frequently cited, is that given by Joseph Scaliger in his *De emendatione temporis*, first published in 1583 and subsequently reprinted many times.[47] Now, if one excludes from Galileo's list the last two intervals, which have already been discussed, and focuses on those between Adam and the birth of Christ, he can see that Galileo's epochs are basically Scaliger's. Actually Scaliger records one more event, the call of Abraham, which he interpolates between Abraham's birth and the Exodus from Egypt, but otherwise he has all the intervals recorded by Galileo, although his figures for them are generally different.

Pererius, on the other hand, has two chronologies, one in his commentary on Daniel and the other in his commentary on Genesis; their difference lies in the fact that they interpolate the reigns of David and of Achaz respectively into the epochs computed.[48] Although Pererius's epochs are thus different from both Scaliger's and Galileo's, with a little computation they can be converted into equivalent epochs, and when this is done it is found that, with the exception of one interval, Pererius's figures are precisely Galileo's. The one exception is the interval between the Exodus from Egypt and the building of the Temple, for which Galileo gives 621 years whereas Pererius, and Scaliger also, give only 480; the difference is that the larger figure (621) is that calculated by the profane historian Josephus, whereas the smaller figure (480) is that computed from the Hebrew text of the Bible.[49] Thus it seems likely that Galileo's figure derives at least indirectly from Pererius and probably comes from a Collegio Romano source like the rest of MS Gal 46.[50]

The question remains as to how such a computation can be reconciled with a composition of the notes 'around 1590,' when the figures themselves seem to indicate 1580, or 1510 years after the destruction of Jerusalem in A.D. 70. A plausible answer is that Galileo made a simple mistake in recording, a mistake that involves only one digit. It seems unlikely that he changed the intervals of the Biblical epochs; whatever his source, he probably took down what was there recorded. But when he came to the interval between the destruction of Jerusalem and 'the present time,' which could be his own computation, he wrote down 1510 instead of 1520, and thus came out ten years short in his entire sum. Now it is not unprecedented that Galileo makes an error of one digit in this way; in fact, in the same MS Gal 46, a few folios

TABLE VI

Possible Sources of the Chronology of Creation in MS Gal 46

Event	Galileo MS Gal. 46	Joseph Scaliger 1583	Pererius (Daniel) 1587	Pererius (Genesis) 1589	Galileo's Chronology Again (1)
Adam					
	1656	1656	1656	1656	1656 (2)
Flood					
	322	292	322	322	322 (3)
Birth of Abraham					
		75			
	505		505		505 (4)
[Call of Abraham]*					
				942	
		430			
Exodus from Egypt					
	621	480	480		621 (5)
[David]					
Building of Temple					
				473	
			283		
[Achaz]	430	427			430 (6)
Captivity of Sedechia					
	70	59			
			776	[630?]	
Dissolution by Cyrus				730	630 (7)
	560	529			
Birth of Christ					
SUBTOTAL	4164	3948	4022	4123 [4023?]	4164 (8)
	74				
Destruction of Jerusalem					
	1510 (9)				
Present					

*Entries enclosed in square brackets are not listed in Galileo's chronology.

TABLE VI

Possible Sources of the Chronology of Creation in MS Gal 46

Notes to Table VI

(1) Galileo's epochs are essentially those of Scaliger.

(2) Same in all sources.

(3) Pererius adds 30 years to Scaliger's figure to allow for the generation of Cain; Galileo follows Pererius.

(4) Same for Scaliger (75 + 430) and for Pererius on Daniel.

(5) Both Scaliger and Pererius on Daniel computer this interval from the Hebrew text (= 480 years); Galileo gives the longer interval found in Josephus (= 621 years) – see William Hales, *New Analysis of Chronology and Geography, History and Prophecy*, (London: 1880), p. 217.

(6) Galileo's figure is consistent with Pererius's two chronologies and can be calculated from them:

$$942 + 473 - 505 - 480 = 430$$

(7) Galileo's figure for this interval is the sum of 70 and 560 or 630; this is about the same as the sum given by Pererius on Daniel:

$$776 + 283 - 430 = 629 \simeq 630$$

It is probable that the figure given by Pererius on Genesis (730) is a misprint and should read 630, as given by Galileo.

(8) Galileo's sum is the same as Pererius's except for one particular, i.e., the interval between the Exodus from Egypt and the building of the Temple – see (5) above; it differs from Scaliger's sum in five particulars, viz, the intervals between the Flood and Abraham, between the Exodus and the Temple, between the Temple and the Captivity, between the Captivity and the Dissolution, and between the Dissolution and the Birth of Christ.

(9) The figure 1510 puts the writing of the notes in A.D. 1580, because the Destruction of Jerusalem took place in A.D. 70, which is 74 years after the Birth of Christ (4 B.C.). If Galileo wrote the wrong figure here, i.e., 1510 instead of 1520, it is necessary to change only one digit to make the date of composition 1590, which agrees with the original notation on the notebook. The date of 1580 for the time of composition would seem to be ruled out by Galileo's extreme youth (16 yrs.) at that time, in light of the sophistication of the notes and their use of materials only available later.

beyond the passage being discussed, when again giving the total number of years from creation to the present time, Galileo there recorded 6748 years, whereas earlier he had given 5748 for that identical sum.[51] There is simply no way of explaining such a different (here a difference of one digit in the thousands column, rather than of one digit in the tens column as just postulated) except ascribing it to an error on Galileo's part. And it would seem much simpler to admit such a slip than to hold that Galileo did the copying in 1580, when he was a mere sixteen years of age, before he had even begun his studies at the University of Pisa.

Moreover, if one persists in the 1580 or 1584 dating, there is no way of accounting for the sophistication of the notes, the lateness of the sources on which they seem to be based, and other historical evidences that will be adduced in the next section. The 1580 dating, which otherwise would be the more plausible, is particularly untenable in light of the dependence of the notes on the 1581 edition of Clavius's *Sphaera*, for, as noted earlier, the copying from Clavius (either directly or through an intermediate) had to be done *after* 1581. On the other hand, if one admits the appreciably later composition consistent with the dates of the *reportationes* of Valla and Vitelleschi (1589 and 1590) and with that of Pererius's commentary on Genesis (1589), then he need no longer view the notes as *Juvenilia*, the writings of a reluctant scholar or of an uncomprehending student, but rather will be disposed to see them as the serious work of an aspiring young professor. From such a viewpoint, of course, the Duhem thesis itself will bear re-examination – not in the form originally suggested by Duhem, to be sure, but in a modified form that takes fuller account of the Collegio Romano materials and their possible influence on Galileo.

VI. MS GAL 46: PROVENANCE AND PURPOSE

Before suggesting such a qualified continuity thesis, it will be helpful to speculate briefly as to how Galileo could have gotten his hands on the sources already discussed, and what use he intended to make of them. In the writer's opinion Clavius remains the key to a proper understanding of these notes. Galileo had met the great Jesuit mathematician during a visit to Rome in 1587, as is known from a letter written by Galileo to Clavius in 1588; thereafter they remained on friendly terms until Clavius's death in 1612.[52] What better explanation for Galileo's possession of all this Collegio Romano material than that it was made available to him by Clavius himself. In Galileo's position at Pisa from 1589 to 1591, and in the openings for a mathematician-astronomer for which Galileo applied at Bologna in 1588 and at Padua in 1592, it would certainly be desirable to have someone with a good knowledge of natural philosophy, who would be conversant, in particular, with matters relating to Aristotle's *De caelo* and *De generatione*. It makes sense to suppose that Clavius would help his young friend, whose mathematical ability he regarded highly, by supplying him with materials from his own Jesuit colleagues that would serve to fill out this desirable background.

Again, among the Jesuits at that time there was a movement to integrate courses in mathematics with those in natural philosophy,[53] and even at Pisa Galileo's predecessor in the chair of mathematics, Filippo Fantoni, had written a treatise *De motu* that was essentially philosophical.[54] Consider this in conjunction with the fact that, both at Pisa and at Padua, professors of philosophy were paid considerably more than professors of mathematics.[55] Could it not be that Galileo was aspiring to a position wherein he could gradually work into philosophy and thus earn for himself a more substantial salary? In this event it would be advantageous for him to have a set of notes, preferably in Latin, that would show his competence in such matters philosophical. What better expedient for him, in these circumstances, than to seek out a good set of lecture notes and from these compose his own. Such composition, in those days, was not regarded as copying and certainly not as plagiarism in the modern sense; it was rather the expected thing, and everyone did it.[56] That would explain Galileo's rearrangement of the arguments, his simplified Latin style, a certain uniformity of presentation not to be found in the varied sources on which the notes seem to be based. And it gives much more meaning to this rather intelligent set of notes, than it does to maintain that they were trite scholastic exercises copied by an uninterested student from a professor whose later publication bears no resemblance to the materials they contain.

Thus far largely internal evidences have been cited that converge toward the year 1590 as the likely time of writing the notes contained in MS Gal 46. Apart from these there are some external evidences confirmatory of this date that may shed some light on the purpose for which the notes were composed. Already mentioned is the dependence of the first set of notes on Clavius's

Sphaera. To this should now be added that, in the second set devoted mainly to *De elementis*, Galileo refers often to the Greek physician Galen, mentioning his name no less than 15 times, and that the *reportatio* of Valla's *Tractatus de elementis* similarly contains extensive references to Galen. Consider then, in relation to these apparently disconnected facts, a letter written by Galileo to his father on November 15, 1590, which reads as follows:

Dear Father: I have at this moment a letter of yours in which you tell me that you are sending me the Galen, the suit, and the *Sfera*, which things I have not as yet received, but may still have them this evening. The Galen does not have to be other than the seven volumes, if that remains all right. I am very well, and applying myself to study and learning from Signor Mazzoni, who sends you his greeting. And not having anything else to say, I close... Your loving son...[57]

The reference to the Galen volumes, in light of what has just been said, could well be significant. Perhaps Galileo, having seen in a secondary source so many citations of Galen, an author with whom he would have been familiar from his medical training, wished to verify some of them for himself.[58] Again, the reference to the *Sfera* is probably not to Sacrobosco's original work but rather to the text as commented on by Clavius. In view of Galileo's personal contact with, and admiration for, the Jesuit mathematician, it would seem that he should turn to the latter's exposition of the *Sphere* rather than to another's. Further, one can suspect from details of Galileo's later controversy with the Jesuits over the comets of 1618 that he was quite familiar with Clavius's commentary.[59] And finally, in Galileo's last published work, the *Two New Sciences* of 1638, he has Sagredo make the intriguing statement, "I remember with particular pleasure having seen this demonstration when I was studying the *Sphere* of Sacrobosco with the aid of a learned commentary."[60] Whose 'learned commentary' would this be, if not that of the celebrated Christopher Clavius?

Again in relation to Clavius, it is known that while at Pisa Galileo taught Euclid's *Elements*, and significantly that he expounded Book V, entitled *De proportionibus*, in the year 1590.[61] This is an unusual book for an introductory course, but it assumes great importance in applied mechanics for treatises such as *De proportionibus motuum*. If Galileo lectured in Latin, it is possible that he used Clavius's commentary on Euclid for this course, for the second edition of that work appeared the year previously, in 1589, and has an extensive treatment of ratios, precisely the subject with which the fifth book is concerned.[62]

Yet again, in 1591 Galileo taught at Pisa the 'hypothesis of the celestial motions,' and then at Padua a course entitled the *Sphera* in 1593, 1599, and 1603, and another course on the *Astronomiae elementa* in 1609.[63] His lecture notes for the Paduan courses have survived in five Italian versions,[64] none of them autographs, and these notes seem to be little more than a popular summary of the main points in Clavius's commentary on Sacrobosco's *Sphaera*. They show little resemblance, on the other hand, to Oronce Finé's commentary on that work, which was used as a text for this course at Pisa as late as 1588.[65] In one of the versions of Galileo's Paduan lecture notes, moreover, there is reproduced a 'Table of Climes According to the Moderns,' lacking in the other versions, but taken verbatim from Clavius's edition of 1581 or later.[66] Indeed, when the *Trattato della Sfera* if compared with the summary of Clavius's *Sphaera* in MS Gal 46, the latter is found to be more sophisticated and richer in technical detail. It could be, therefore, that the notes contained in MS Gal 46 represent Galileo's first attempt at class preparation, and that later courses based on the *Sphaera* degenerated with repeated teaching – a phenomenon not unprecedented in professorial ranks.

Another possible confirmation is Galileo's remark in the first essay version of the *De motu antiquiora* to the effect that his "commentaries on the *Almagest* of Ptolemy. . .will be published in a short time."[67] In these commentaries, he says, he explains why objects immersed in a vessel of water appear larger than when viewed directly. No commentary of Galileo on Ptolemy has yet been found, but significantly this particular phenomenon is discussed in Clavius's commentary on the *Sphere*.[68] The latter also effectively epitomizes the *Almagest* and so could well be the source Galileo had in mind for his projected summary.

All of this evidence, therefore, gives credence to the thesis emerging out of this study, namely, that the materials recorded in MS Gal 46 need no longer be regarded simply as *Juvenilia*, as the exercises of a beginning student; perhaps more plausibly can they to be seen as lecture notes or other evidences of scholarship composed by Galileo in connection with his Pisan professorship from 1589 to 1591.[69] As such they then merit serious consideration, not merely for the insight they furnish into Galileo's intellectual formation, but for identifying the philosophy with which he operated during the first stages of his teaching career.

VII. CONTINUITY REVISITED: GALILEO AND THE *PARISIENSES*

Now back to the thorny problem of continuity. Favaro, as is known from his critique, did not possess the facts here presented; so impressed was he by the novelty of Galileo's contribution, moreover, that he was not disposed to discern any law of continuity operating on the thought of his master. Duhem, at the other extreme, was disposed to see continuity even where none existed. Duhem's own philosophy of science was decidedly positivistic, placing great emphasis on "saving the appearances," according no realist value to scientific theories.[70] In his eyes the nominalists of the fourteenth century, the *Doctores Parisienses*, had the correct view, and he wanted to connect Galileo directly with them. The afore-mentioned evidences, unfortunately for Duhem, will not support such an immediate relationship.[71] And yet they do suggest some connection of early modern science with medieval science *via* the writings of sixteenth-century authors, along lines that will now be sketched.

In his studies on Leonardo da Vinci, Duhem correctly traced nominalist influences from the *Doctores Parisienses* all the way to Domingo de Soto, the Spanish Dominican who taught at Salamanca in the 1540's and 1550's.[72] Now it is an interesting fact, in connection with Soto, that many of the early professors at the Collegio Romano were either Spanish Jesuits, such as Pererius, or they were Jesuits of other nationalities who had completed some of their studies in Spain or Portugal, such as Clavius, who had studied mathematics at Coimbra under Pedro Nuñez. Warm relationships also existed between Pedro de Soto (a blood relative of Domingo and also a Dominican) and the Roman Jesuits, as Villoslada notes in his history of the Collegio Romano.[73] Admittedly, Duhem pursued his theme a long way, but as it has turned out he did not pursue it far enough. In the passage from fourteenth-century Paris to sixteenth-century Spain, moreover, Duhem failed to note that many nominalist theses had given way to moderate realist theses, even though the techniques of the *calculatores* continued to be used in their exposition. Duhem never did study the further development, the passage from Spain to Italy, when other changes in methodological orientation took place particularly among the Jesuits. The nominalist emphasis on the logic of consequences, for example, quickly ceded to a realist interest in Aristotle's *Posterior Analytics* and in the methodology explained therein for both physical and physico-mathematical sciences. The Averroist atmosphere in

Italian universities also led to other emphases, to a consideration of problems different from those discussed in fourteenth-century Paris. The movement that gave rise to Galileo's science, therefore, may well have had its origin, its *terminus a quo*, among the *Doctores Parisienses*, but its *terminus ad quem* can be called nominalist only in a much attenuated sense. More accurately can it be described as an eclectic scholastic Aristotelianism deriving predominantly from Aquinas, Scotus, and Averroës, and considerably less from Ockham and Buridan, although it was broadly enough based to accomodate even the thought of such nominalist writers.

Exemplification of this qualified continuity thesis can be given in terms of Galileo's treatment of topics in natural philosophy and in logic that are usually said to have nominalist connotations. Among the former could be listed, for example, creation, the shapes of the elements, maxima and minima, uniformly difform motion, impetus, the intension and remission of forms, degrees of qualities, and analyses of infinities. For purposes here only one such topic need be discussed, viz., the reality of local motion, for this can serve to illustrate some of the conceptual changes that took place in the centuries between Ockham and Galileo.

For William of Ockham, as is well known, local motion was not an entity in its own right but could be identified simply with the object moved. To account for local motion, therefore, all one need do is have recourse to the moving body and its successive states; the phenomenon, as Ockham said, "can be saved by the fact that a body is in distinct places successively, and not at rest in any."[74] Since this is so it is not necessary to search for any cause or proximate mover in the case of local motion: "local motion is not a new effect. . .it is nothing but that a mobile coexist in different parts of space."[75]

In contrast to Ockham, as is also well known, Jean Buridan subscribed to the traditional analysis of local motion, with consequences that were quite significant for the history of science. In his commentary on Aristotle's *Physics* Buridan inquired whether local motion is really distinct from the object moved, and answered that while the "later moderns" (*posteriores moderni* – an obvious reference to Ockham) hold that it is not, he himself holds that it is, and went on to justify his conclusion with six different arguments.[76] Thus, for Buridan and other Parisians, local motion was decidedly a "new effect"; it was this conviction that led them to study it *quoad causes* and *quoad effectus* and, as part of the former investigation, to develop their theories of impetus, the proximate forerunner of the modern concept of inertia.

As has been indicated elsewhere, a development similar to Buridan's among sixteenth-century commentators such as Juan de Celaya and Domingo de Soto provided the moderate realist background against which Galileo's *De motu antiquiora* must be understood.[77] Soto was not adverse to calculatory techniques; in fact he was the first to apply them consistently to real motions found in nature, such as that of free fall, and so could adumbrate Galileo's 'law of falling bodies' some 80 years before the *Two Chief World Systems* appeared.[78] Moreover, Jesuits such as Clavius saw the value of applying mathematical techniques to the study of the world of nature. In a paper written for the Society of Jesus justifying courses of mathematics in Jesuit *studia*, Clavius argued that without mathematics "physics cannot be correctly understood," particularly not matters relating to astronomy, to the structure of the continuum, to meteorological phenomena such as the rainbow, and to "the ratios of motions, qualities, actions, and reactions, on which topics the *calculatores* have written much."[79] And it is significant that the *reportatio* of Vitelleschi discussed above concludes with the words: "And thus much concerning the elements, for matters that pertain to their shape and size partly have been explained by us elsewhere, partly are presupposed from the *Sfera*"[80] – an indication that by his time at the Collegio the mathematics course was already a prerequisite to the lectures on the *De caelo* and *De generatione*.

Duhem's thesis with regard to the study of motion, therefore, requires considerable modification, and this along lines that would accent realist, as opposed to nominalist, thought. With regard to logic and methodology, on the other hand, his thesis runs into more serious obstacles. Although the nominalist logic of consequences was known to the Jesuits, and Bellarmine preferred to express himself in its terms in his letter to Foscarini,[81] it was never adopted by them as a scientific methodology, Clavius, in particular, ruled it out as productive of true science in astronomy. So he argued that if one were consistently to apply the principle *ex falso sequitur verum*,

> then the whole of natural philosophy is doomed. For in the same way, whenever someone draws a conclusion from an observed effect, I shall say, "that is not really its cause; it is not true because a true conclusion can be drawn from a false premise." And so all the natural principles discovered by philosophers will be destroyed.[82]

Galileo followed Clavius's causal methodology in his own attempted proof for the Copernican system based on the tides, which was never viewed by him as merely "saving the appearances." He also used the canons of the *Posterior Analytics* in his final work, the *Two New Sciences*, wherein he proposed to found a new mixed science of local motion utilizing a method of demonstrating *ex suppositione*. This methodological aspect of Galileo's work has been examined at greater length by Crombie and by the writer in recent publications, to which the reader is referred for fuller details.[83] Here a few supplementary observations may suffice to indicate the main thrust of the argument.

Ockham had a theory of demonstration just as he had a theory of motion, but for him a demonstration is nothing more than a disguised hypothetical argumentation, thus not completely apodictic – a typical nominalist position.[84] Jean Buridan, here as in the case of the reality of motion, combatted Ockham's analysis and returned to an earlier position expounded by Thomas Aquinas, showing that in natural science, and likewise in ethics, truth and certitude can be attained through demonstration, but it must be done by reasoning *ex suppositione*.[85] Now Galileo, in this matter, turns out to be clearly in the tradition of Aquinas and Buridan, not in that of Ockham. He in fact mentions the procedure of reasoning *ex suppositione* at least six times in the writings published in the National Edition: once in 1615 when arguing against Bellarmine's interpretation that all *ex suppositione* reasoning must be merely hypothetical, once in the *Two Chief World Systems*, twice in the *Two New Sciences*, and twice in correspondence explaining the methodology employed in the last named work.[86] In practically every one of these uses Galileo gives the expression a demonstrative, as opposed to a merely dialectical, interpretation. Others may have equated suppositional reasoning with hypothetical reasoning, but Galileo consistently accorded it a more privileged status, seeing it as capable of generating true scientific knowledge.

Moreover, it seems less than coincidental that Galileo's earliest use of the expression *ex suppositione* occurs in his series of logical questions, the notes on Aristotle's *Posterior Analytics* referred to above, which otherwise are very similar to the physical questions that have been the focal point of this study. There, in answering the query whether all demonstration must be based on principles that are "immediate" (i.e., *principia immediata*), Galileo replies to the objection that mixed or "subalternated sciences have perfect demonstrations" even though not based on such principles, as follows:

I answer that a subalternated science, being imperfect, does not have perfect demonstrations, since it supposes first principles proved in a superior [science]; therefore it generates a science *ex suppositione* and *secundum quid.*[87]

Now compare this terminology with a similar sentence from Pererius, who, in a *reportatio* of his lectures on the *De caelo* given at the Collegio Romano around 1566, states:

The *Theorica planetarum*. . .either is a science *secundum quid* by way of *suppositio* or it is merely opinion, not indeed by reason of its consequent but by reason of its antecedent.[88]

Note that both Galileo and Pererius allow the possibility of science being generated from reasoning *ex suppositione*, and that Pererius explicitly distinguishes a science so generated from mere opinion. It thus seems far from unlikely that Galileo adopted the Jesuit ideal of a mixed or subalternated science (the paradigm being Clavius's mathematical physics) as his own, and later proceeded to develop his justification of the Copernican system, and ultimately his own 'new science' of motion, under its basic inspiration.[89] On this interpretation Galileo's logical methodology would turn out to be initially that of the Collegio Romano, just as would his natural philosophy – and this is not nominalism, but the moderate realism of the scholastic Aristotelian traditicn.[90]

Viewed from the perspective of this study, therefore, nominalism and the *Doctores Parisienses* had little to do proximately with Galileo's natural philosophy or with his methodology.[91] This is not to say that either the movement or the men were unimportant, or that they had nothing to contribute to the rise of modern science. Indeed, they turn out to be an important initial component in the qualified continuity thesis here proposed, chiefly for their development of calculatory techniques that permitted the importation of mathematical analyses into studies of local motion, and for their promoting a "critical temper," to use John Murdoch's expression, that made these and other innovations possible within an otherwise conservative Aristotelianism.[92] But Galileo was not the immediate beneficiary of such innovations; they reached him through other hands, and incorporated into a different philosophy. What in fact probably happened is that the young Galileo made his own the basic philosophical stance of Clavius and his Jesuit colleagues at the Collegio Romano, who had imported nominalist and calculatory techniques into a scholastic Aristotelian synthesis based somewhat eclectically on Thomism, Scotism, and Averroism. To these, as is well known, Galileo himself added Archimedean and Platonic elements, but in doing so he remained committed to Clavius's realist ideal of a mathematical physics that demonstrates truth about the physical universe. And Jesuit influences aside, there can be little doubt that Galileo consistently sided with realism, as over against nominalism, in physics as in astronomy, from his earliest writings to his *Two New Sciences*. At no time, it would appear, did he subscribe to Duhem's ideal of science that at best attains only hypothetical results and at worst merely "saves the appearances."

What then is to be said of Favaro's critique of the Duhem thesis? An impressive piece of work, marred only by the fact that Favaro did not go far enough in his historical research, and thus lacked the materials on which a nuanced account of continuity could be based. As for Duhem's 'precursors,' they surely were there, yet not the precise ones Duhem had in mind, nor did they think in the context of a philosophy he personally would have endorsed. But these defects notwithstanding, Favaro and Duhem were still giants in the history of science. Without their efforts we would have little precise knowledge of either Galileo or the *Doctores Parisienses*, let alone the quite complex relationships that probably existed between them.

The Catholic University of America,
Washington, D.C.

NOTES

* Research on which this paper is based has been supported by the National Science Foundation (Grant No. SOC 75-14615), whose assistance is gratefully acknowledged. Apart from the earlier version presented at the Blacksburg Workshop on Galileo on October 26, 1975, portions of this essay have been presented at the Folger Shakespeare Library in Washington, D.C., on October 20, 1975, and at meetings of the American Philosophical Association in New York on December 28, 1975, and of the History of Science Society in Atlanta on December 30, 1975. The author wishes particularly to acknowledge the comments, among many others, of Alistair Crombie, Stillman Drake, William F. Edwards, Paul O. Kristeller, Charles H. Lohr, Michael S. Mahoney, John E. Murdoch, Charles B. Schmitt, and Thomas B. Settle, which have aided him appreciably in successive revisions.

[1] Galileo Galilei e i Doctores Parisienses', *Rendiconti della R. Accademia dei Lincei* 27 (1918), 3-14.

[2] Léonard da Vinci a-t-il exercé une influence sur Galilée et son école?' *Scientia* 20

(1916), 257–265.

[3] The manuscript referred to here is contained in the Biblioteca Nazionale Centrale in Florence with the signature Manoscritti Galileiani 46 (= BNF/MS Gal 46). Its transcription is contained in Antonio Favaro (ed.), *Le Opere di Galileo Galilei*, 20 vols. in 21, G. Barbera Editore, Florence, 1890–1909, reprinted 1968, Vol. 1, pp. 15–177 (hereafter abbreviated as *Opere* 1:15–177). All manuscript material in this essay is reproduced with the kind permission of the Biblioteca Nazionale Centrale in Florence, Italy. This permission is here gratefully acknowledged.

[4] *Études sur Léonard de Vinci*, 3 vols., A. Hermann & Fils, Paris, 1913, Vol. 3, p. 583.

[5] "...che lo sviluppo scientifico e soggetto alla legge di continuità; che le grandi scoperte sono quasi sempre il frutto d'una preparazione lenta e complicatà, proseguita attraverso i secoli; che in fine le dottrine, le quali i più insigni pensatori giunsero a professare, risultano da una moltitudine di sforzi accumulati da una folla di oscuri lavoratori." – 'Galileo Galilei e i Doctores Parisienses,' p. 4 (translation here and hereafter by the writer).

[6] "Galileo Galilei e i Doctores Parisienses,' pp. 8–10.

[7] "Senza dubbio alcuno, dunque, ciò scriveva Galileo durante l'anno 1584." – Avvertimento, *Opere* 1:12.

[8] "Galileo Galilei e i Doctores Parisienses,' p. 10; cf. Avvertimento, *Opere* 1:12. The reference is to Francesco Buonamici, *De motu libri decem*...Apud Bartholomaeum Sermatellium, Florentiae, 1591.

[9] 'Galileo Galilei e i Doctores Parisienses,' p. 11; see also *Opere* 1:12–13 and 9:275–282.

[10] Favaro reports this inscription in his Avvertimento, *Opere* 1:9, as "L'esame dell'opera d'Aristotele 'De Caelo' fatto da Galileo circa l'anno 1590."

[11] The complete list of questions, as given by Galileo in MS Gal 46, is as follows:

Quaestio prima. Quid sit id de quo disputat Aristoteles in his libris De caelo.
Quaestio secunda. De ordine, connexione, et inscriptione horum librorum.

Tractatio prima. De mundo.
Quaestio prima. De opinionibus veterum philosophorum de mundo.
Quaestio secunda. Quid sentiendum sit de origine mundi secundum veritatem.
Quaestio tertia. De unitate mundi et perfectione.
Quaestio quarta. An mundus potuerit esse ab aeterno.

Tractatio de caelo.
Quaestio prima. An unum tantum sit caelum.
Quaestio secunda. De ordine orbium caelestium.
Quaestio tertia. An caeli sint unum ex corporibus simplicibus, vel ex simplicibus compositi.
Quaestio quarta. An caelum sit incorruptibile.
Quaestio quinta. An caelum sit compositum ex materia et forma.
Quaestio sexta. An caelum sit animatum.

[Tractatus de alteratione]
[Quaestio prima. De alteratione.]
Quaestio secunda. De intensione et remissione.
Quaestio ultima. De partibus sive gradibus qualitatis.

Tractatus de elementis
Prima pars. De quidditate et substantia elementorum.

Prima quaestio. De definitionibus elementi.
Quaestio secunda. De causa materiali, efficiente, et finali elementorum.
Quaestio tertia. Quae sint formae elementorum.
Quaestio quarta. An formae elementorum intendantur et remittantur.
Secunda disputatio. De primis qualitatibus.
Quaestio prima. De numero primarum qualitatum.
Quaestio secunda. An omnes hae quatuor qualitates sint positivae, an potius aliquae sint privativae.
Quaestio tertia. An omnes quatuor qualitates sint activae.
Quaestio quarta. Quomodo se habeant primae qualitates in activitate et resistentia.

Note in the above list that Galileo is not consistent in listing the divisions, i.e., that the first two tractates are given with the Latin title *Tractatio*, the last two with the title *Tractatus*, and that the last tractate is subdivided into a *Prima pars* and a *Secunda disputatio* (themselves inconsistent divisions), whereas the previous tractates are subdivided directly into *Quaestiones*. Such inconsistencies could be signs of copying from several different sources; see note 35 *infra*.

[12] Thus, in *Opere* 1, at p. 122, line 10 (hereafter abbreviated as 122.10), Galileo makes the reference to "[ea] quae dicta sunt a nobis 6° Physicorum. . .," an indication of a commentary on the eight books of the *Physics*, and at 137.14 he speaks of difficulties that will be solved "cum agam de elementis in particulari," a common way of designating the subject matter of the *Meteorology*, of which he apparently planned to treat. There is also an implicit reference to notes on logic that probably preceded the notes on natural philosophy; this occurs at 18.17, where Galileo writes ". . .de singularibus non potest esse scientia, ut alibi ostendimus," a possible indication of his having already explained this thesis from the *Posterior Analytics*. Other references to matter treated elsewhere occur at 77.18, 77.24, 113.13, 125.25–32, 127.30, 128.8, 129.24, 138.9, and 150.24–25.

[13] All of these references to Aquinas and the Thomists have been analyzed by the writer in W. A. Wallace, 'Galileo and the Thomists', *St. Thomas Aquinas Commemorative Studies 1274–1974* (ed. by A. Maurer), Pontifical Institute of Mediaeval Studies, Toronto, 1974, Vol. 2, pp. 293–330. At the time he wrote that article the writer accepted uncritically Favaro's judgment that the notes contained in MS Gal 46 were *Juvenilia*; now, as will become clear in what follows, he seriously questions such a characterization of the notes.

[14] For an account of the discoveries of Crombie and Carugo, see A. C. Crombie, 'Sources of Galileo's Early Natural Philosophy', in *Reason, Experiment, and Mysticism in the Scientific Revolution* (ed. by M. L. Righini Bonelli and W. R. Shea) Science History Publications, New York, 1975, pp. 157–175, 303–305. For the results of the author's researches, apart from those reported in note 13 above, see W. A. Wallace, 'Galileo and Reasoning *Ex Suppositione*: The Methodology of the *Two New Sciences*,' in Boston Studies in the Philosophy of Science, Vol. XXXII (Proceedings of the 1974 Biennial Meeting of the Philosophy of Science Association, 1974) (ed. by R. S. Cohen *et al.*), Reidel, Dordrecht and Boston, 1976, pp. 79–104; note 3a of this paper, added in proof, clarifies matters relating to priorities of discovery.

[15] The Latin text, also given below, is in *Opere* 1:38.4–9.

[16] The relevant folios of MS Gal 46 are fol. 16v to fol. 26r, which duplicate matter found on pp. 42–46, 55–71, and 135–143 of the 1581 edition of Clavius's *Sphaera*.

[17] This explains Galileo's apparent skipping of extensive passages in Clavius's text, viz., those on pp. 47–54 and 73–134.
[18] The ten folios of Galileo's notes contain about 5650 words, as compared with some twelve thousand words in the related passages of Clavius.
[19] Galileo himself alludes to this fact in his third letter on sunspots to Mark Welser, dated December 1, 1612; see *Opere* 5:190.2–7.
[20] Galileo could have used the 1570 edition for portions of his work, but he incorporates material not found in that edition, and for this he must have used (directly or indirectly) either the 1581 or the 1585 edition, both of which were printed from the same type. For a complete listing of Clavius's writings, see Carlos Sommervogel, S.J., *Bibliothèque de la Compagnie de Jésus*, Vol. II (Alphonse Picard, Paris, 1891) cols. 1212–1224; see also note 33 *infra.*
[21] It should be noted here, however, that other pieces of evidence count against this interpretation, and these will be discussed below in conjunction with the investigation of possible manuscript sources of MS Gal 46 (see text corresponding to notes 28 and 34, *infra*). Also noteworthy is Sommervogel's indication that Clavius's commentary was abridged in a printed edition published at Cologne in 1590 (see his Tom. II, col. 1213); the writer has not yet located this abridgment, which could of course be the source of Galileo's note-taking. If so, however, this would definitively date the composition of MS Gal 46 in 1590 or later.
[22] For the Latin text, see *Opere* 1:24.25–27. The reference to Pererius is apparently to his widely available *De communibus omnium rerum naturalium principiis et affectionibus libri quindecim*, printed at Rome in 1576 and often thereafter. It should be noted, however, that Sommervogel (Vol. VI, col. 499) lists an earlier work printed at Rome in 1562 with the title *Physicorum, sive de principiis rerum naturalium libri XV*; this is much rarer, and I have not yet been able to locate a copy. There are also manuscript versions of Pererius's lectures on the *Physics*, as noted *infra*, p. 107.
[23] Galileo's reply is on fol. 14v; Pererius's corresponding discussion is on p. 505 of the 1576 edition.
[24] For details, see R. G. Villoslada, *Storia del Collegio Romano dal suo inizio (1551) alla soppressione della Compagnia di Gesù (1773)*. Analecta Gregoriana Vol. LXVI, Gregorian University Press, Rome, 1954, especially pp. 89–91 and 329–335.
[25] The manuscript is in Rome in the Archivum Pontificiae Universitatis Gregorianae, Fondo Curia, MS 392 (= APUG/FC 392).
[26] Material enclosed in curly parentheses { } has been rearranged to show parallels in the texts.
[27] Note that for Pererius, as shown on p. 102 *supra*, the opinion of Philoponus and Henry of Ghent is his first, whereas for Galileo and Vitelleschi it is their third.
[28] This is one instance counting against the *prima facie* evidence for Galileo's copying from printed sources, alluded to above (note 21).
[29] For the Latin text, which is reproduced in part below, see *Opere* 1:138.3–9.
[30] The last folio of APUG/FC 392 concludes with the sentence: "...Et hec de elementis, nam quae spectant ad eorum figuram et quantitatem partim explicata sunt a nobis alibi, partim supponimus ex *Sfera*."
[31] In a private communication to the author, Stillman Drake has pointed out that Galileo could have had in mind the material he drafted for the *De motu antiquiora*, as in *Opere* 1:345–346, which gives such proofs and likewise has a marginal reference to "p^{o}

Meteororum cap. 3°." This would establish a hitherto unnoticed connection between MSS 46 and 71.

[32] On the other hand, as Charles Lohr has indicated to me, the peculiar manner of citation, and the fact that Marsilius of Inghen is mentioned by Galileo whereas Buridan, Albert of Saxony, and Themo Judaeus are not (see Table I), suggests some dependence, possibly indirect, on a compilation such as Lokert's Paris edition of 1516, as Duhem originally speculated.

[33] The earliest edition recorded by Sommervogel (Vol. II, col. 1212) is a quarto volume published at Rome by Victorius Helianus in 1570. Some caution is necessary here, however, as the writer has seen reference to a quarto edition of this work published at Rome in 1565. This is in a handwritten inventory of books that at one time were in the personal library of Pope Clement CI, many of which are now in the Clementine Collection of the Catholic University of America, Washington, D.C.; unfortunately this particular Clavius volume is no longer in the collection.

[34] This is a second piece of evidence counting against Galileo's having copied directly from a printed source; see note 21 above.

[35] As remarked at the end of note 11, the inconsistencies of titling could be a sign of reliance on different sources. In this connection it is perhaps noteworthy that the primary division of Vitelleschi's work is the *Tractatio*, which is subdivided into the *quaestio*, as in Galileo's first set of notes; the primary division of Valla's work, on the other hand, is the *Tractatus*, and this is subdivided into the *disputatio* and the *pars*, and then finally into the *questio*, as in Galileo's second set of notes.

[36] See Favaro's Avvertimento in *Opere* 9:279–282.

[37] See Favaro's Avvertimento in *Opere* 9:275–276, together with his transcription of the text and an indication of the corrections, *ibid.*, 9:283–284.

[38] For the Latin text, see *Opere* 1:27.

[39] See Favaro's Avvertimento in *Opere* 1:9 and 1:11–12; also 'Galileo Galilei e i Doctores Parisienses,' p. 8.

[40] This date was surely known in the time of Galileo; it is recorded, for example, by Joseph Scaliger in his widely used *De emendatione tempiris*, first printed in 1583, as taking place in A.D. 70.

[41] Rev. William Hales, D.D., *A New Analysis of Chronology and Geography, History and Prophecy*. . ., 4 vols., London: C. J. G. & F. Rivington, 1830, Vol. 1, p. 214.

[42] *Ibid.*, pp. 211–214.

[43] In his researches the writer has uncovered only one book that does give Galileo's figure. This is Ignatius Hyacinthus Amat de Graveson, O.P., *Tractatus de vita, mysteriis, et annis Jesu Christi*. . ., Venetiis: Apud Joannem Baptistam Recurti, 1727, pp. 251–252: "Christus Dominus anno aerae vulgaris vigesimo sexto, imperii proconsularis Tiberii decimo sexto, anno urbis Romae conditae 779, anno a creatione mundi 4164. . ." The date of publication of this work obviously would rule it out as a source; but see note 50 *infra*.

[44] For details concerning Pererius's Averroism and internal controversies at the Collegio, see M. Scaduto, *Storia della Compagnia di Gesù in Italia*. L'Epoca di Giacomo Lainez, 2 vols., Gregorian University Press, Rome, 1964, Vol. 2, p. 284. Also R. G. Villoslada, *Storia del Collegio Romano*, pp. 52, 78ff, 329, and C. H. Lohr, 'Jesuit Aristotelianism and Suarez's *Disputationes Metaphysicae*,' to appear in *Paradosis: Studies in Memory of E. A. Quain* (New York 1976).

[45] Villoslada, *Storia del Collegio Romano*, p. 323.
[46] Benedictus Pererius, *Commentariorum in Danielem prophetam libri sexdecim...* Romae: Apud Georgium Ferrarium, 1587; and *Prior tomus commentariorum et disputationum in Genesim...* Romae: Apud Georgium Ferrarium, 1589.
[47] We have used the Frankfurt 1593 edition, where the chronology is given on p. 377.
[48] For details, see the chronologies listed by Pererius on pp. 350–351 of his commentary on Daniel and on p. 130 of his commentary on Genesis.
[49] See Hales, *A New Analysis of Chronology*, p. 217.
[50] It is perhaps significant that the author cited in note 42, Amat de Graveson, taught at the Collegio Casanatense in Rome, which was adjacent to the Collegio Romano, and thus he could have had access to its archives.
[51] It should be noted, moreover, that the 5748-year total on fol. 10r was written with some hesitation: first Galileo wrote 50 in Arabic numerals, then crossed out these numbers, then wrote 'five thousand four and eighty' in longhand, then crossed out the 'four and eighty,' changed it to 'eight and forty,' and finally inserted a 'seventy' before the 'eight and forty,' all in longhand. Now none of the other figures on this folio were changed in any way; they all appear exactly as in the translation on p. 114. On fol. 15v, however, when recording what should have been the same number of years, Galileo did not give the same total but wrote instead 'six thousand seven hundred and 48 years,' without apparently noticing the difference of a thousand years. The successive revisions in the first sum could be an indication that this was Galileo's own calculation rather than something he copied from an existing source, the result of which calculation he had difficulty putting into Latin prose.
[52] See *Opere* 10:22; also Villoslada, *Storia del Collegio Romano*, pp. 194–199.
[53] For details, see Giuseppe Cosentino, 'L'Insegnamento delle Matematiche nei Collegi Gesuitici nell'Italia settentrionale. Nota Introduttiva', *Physis* 13 (1971), 205–217, and 'Le mathematiche nella "Ratio Studiorum" della Compagnia di Gesù', *Miscellanea Storica Ligure* (Istituto di Storia Moderna e Contemporanea, Università di Genova), II, 2 (1970), 171–213.
[54] See Charles B. Schmitt, 'The Faculty of Arts at Pisa at the Time of Galileo', *Physis* 14 (1972), 243–272, especially p. 260.
[55] Schmitt, 'The Faculty of Arts...,' p. 256; also Ludovico Geymonat, *Galileo Galilei* (transl. by Stillman Drake), McGraw-Hill Book Co., New York, 1965, pp. 10–11.
[56] Galileo's friend, Mario Guiducci, makes this point very well in his Letter to Father Tarquinio Galuzzi, where he states: "It seems to me that...it is wrong to call men copyists who, when treating a philosophical question, take an idea from one author or another and, as is not the case with those who merely copy the writings, make it their own by judiciously adapting it to their purposes so as to prove or disprove some or other statement....To give an exceptional example, on these terms Father Christopher Clavius would have been a first-class copyist, for he was extremely diligent in extracting and compiling in his works of great erudition the opinions and demonstrations of the most distinguished geometers and astronomers up to his time – as seen in his compendius commentary on the *Sphere* of Sacrobosco and in so many other of his works." – *Opere* 6:189; see note 59 *infra*.
[57] *Opere* 10:44–45. This letter and its contents are discussed by Crombie in his paper cited in note 14 *supra*, pp. 167–68.
[58] This possibility, of course, reopens the question as to whether or not Galileo actually

did have recourse to primary sources, in some instances at least, when composing these notes; the mere fact that the notes are based on secondary sources, or bear close resemblances to existing manuscripts, does not preclude consultation of the originals cited therein.

[59] Favaro speculates that Galileo might have had a hand in Guiducci's Letter, cited in note 56 *supra*; see *Opere* 6:6. For other collaboration between Galileo and Guiducci, see William R. Shea, *Galileo's Intellectual Revolution.* Middle Period, 1610–1632, Science History Publications, New York, 1972, pp. 75–76.

[60] *Opere* 8:101.

[61] See Schmitt, 'The Faculty of Arts. . .,' pp. 261–262.

[62] Christophorus Clavius, *Euclidis Elementorum Libri XV*. . ., Romae: Apud Vincentium Accoltum, 1574; nunc iterum editi ac multarum rerum accessione locupletati, Romae: Apud Bartholomaeum Grassium, 1589. Galileo's own interpretation of Euclid, however, would still derive from Tartaglia's Italian translation, with which he was quite familiar, and which, as Stillman Drake has repeatedly argued, underlies his distinctive geometrical approach to the science of motion.

[63] Schmitt, 'The Faculty of Arts. . .,' p. 262; also *Opere* 19:119-120; and Antonio Favaro, *Galileo Galilei a Padova*, Padua: Editrice Antenore, 1968, pp. 105–114, especially p. 108.

[64] Four versions are listed by Favaro, *Opere* 2:206; Stillman Drake reports a fifth version, 'An Unrecorded Manuscript Copy of Galileo's *Cosmography*', *Physis* 1 (1959), 294–306.

[65] See Schmitt, 'The Faculty of Arts. . .,' p. 206.

[66] The table is reproduced in *Opere* 2:244–245; compare this with the table on pp. 429–430 of the 1581 edition of Clavius's *Sphaera.*

[67] *Opere* 1:314.

[68] Clavius, *Sphaera* (1581 edition), pp. 108–109.

[69] Crombie has suggested a similar re-evaluation of the so-called *Juvenilia* in his 'Sources. . .' Paper (note 14 above), pp. 162–170. In this connection it is not essential to the thesis here proposed that the notes actually have been written in 1590 – the date toward which the evidences adduced above appear to converge. If the dependence on the Collegio Romano materials be conceded, there are several possibilities for the transmission of these materials to Galileo. Galileo could have obtained them from Clavius as early as 1587, for use when tutoring in Florence and Siena or for securing the vacant teaching posts at Bologna or at Pisa. Again, he may have obtained them independently of Clavius, and thus at an even earlier date. For example, I have discovered that notes very similar to those of the Collegio were used by a Benedictine monk at the University of Perugia in 1590; it is quite possible that such notes were disseminated throughout other monasteries and religious orders in northern Italy. Now Galileo spent some time as a youth at the monastery of Vallombrosa and remained on friendly terms with the monks there, even teaching a course on the *Perspectiva* for them in 1588 (the latter information *via* a personal communication to the author from Thomas B. Settle). In such a setting he could have had access to *reportationes* of the type described and have made his own notes from them so as to be prepared for an eventual teaching assignment. The fact remains, however, that the materials discussed above are the only such *reportationes* that have been discovered thus far, and in defect of other evidences the 1590 dating alone has more than conjectural support.

[70] For Duhem's own views see his *The Aim and Structure of Physical Theory* (transl. by P. P. Wiener), The Princeton University Press, Princeton, 1954; for their historical justification, apart from Duhem's monumental work on *Le Système du Monde*, see his *To Save the Phenomena*, 'An Essay on the Idea of Physical Theory from Plato to Galileo' (transl. by E. Doland and C. Maschler), University of Chicago Press, Chicago, 1969.
[71] There are other flaws in Duhem's historical arguments, of course, and these have been well detailed by Annaliese Maier in her studies on the natural philosophy of the late scholastics. Other scholars have contributed substantial information, since Duhem's time, that support various aspects of his continuity thesis; among these should be mentioned Ernest A. Moody and Marshall Clagett and their disciples. In what follows, to the researches of these authors will be added a brief survey of sixteenth-century work that complements their findings but leads to slightly different philosophical conclusions than have heretofore been argued.
[72] The last half of Duhem's third volume on Leonardo da Vinci is in fact entitled 'Dominique Soto et la Scolastique Parisienne,' pp. 263–581, of which pp. 555–562 are devoted to Soto's teachings. For a summary of Soto's life and works, with bibliography, see the article on him by the writer in the *Dictionary of Scientific Biography*, Vol. 12, Charles Scribner's Sons, New York, 1975, pp. 547–548.
[73] *Storia del Collegio Romano*, pp. 60–61. Villoslada also calls attention to the professed Thomism of the theology faculty there, and to the tendency otherwise to imitate the academic styles then current at the Universities of Paris and Salamanca (p. 113).
[74] Cited by Herman Shapiro, *Motion, Time and Place According to William Ockham*, Franciscan Institute Publications, St. Bonaventure, N.Y., 1957, p. 40.
[75] *Ibid.*, p. 53; see also William of Ockham, *Philosophical Writings: A selection* (ed. by Philotheus Boehner), Thomas Nelson & Sons, Ltd., Edinburgh, 1957, p. 156.
[76] *Subtilissime questiones super octo phisicorum libros Aristotelis*, Lib. 3, q. 7, Parisiis: In edibus Dionisii Roce, 1509, fols. 50r–51r.
[77] W. A. Wallace, 'The Concept of Motion in the Sixteenth Century', *Proceedings of the American Catholic Philosophical Association* 41 (1967), 184–195; and 'The "Calculatores" in Early Sixteenth-Century Physics', cited *supra* in note 14.
[78] W. A. Wallace, 'The Enigma of Domingo de Soto: *Uniformiter difformis* and Falling Bodies in Late Medieval Physics', *Isis* 59 (1968), 384–401.
[79] The text of that paper, written around 1586, is cited by Cosentino, 'Le matematiche nella "Ratio Studiorum"...,' p. 203, as follows: "Senza le matematichi 'physicam... recte percipi non potest, praesertim quod ad illam partem attinet, ubi agitur de numero et motu orbium coelestium, de multitudine intelligentiarum, de effectibus astrorum, qui pendent ex variis coniunctionibus, oppositionibus et reliquis distantiis inter sese, de divisione quantitatis continuae in infinitum, de fluxu et refluxu maris, de ventis, de cometis, iride, halone et aliis rebus meteorologicis, de proportione motuum, qualitatum, actionum, passionum et reactionum, etc., de quibus multa scribunt Calculatores.'"
[80] Latin text cited above, note 30.
[81] In this letter, dated April 12, 1615, Bellarmine commended Foscarini and Galileo for being prudent in contenting themselves to speak hypothetically and not absolutely when presenting the Copernican system, thus considering it merely as a mathematical hypothesis (*Opere* 12:171–172). Galileo, of course, quickly disavowed that such was his

intent (*Opere* 5:349–370, especially p. 360).

[82] Latin text in Clavius, *Sphaera*, 1581 edition, p. 605. An English translation of this and surrounding passages is to be found in R. Harré, *The Philosophies of Science: An Introductory Survey*, Oxford University Press, Oxford, 1972, pp. 84–86.

[83] Crombie, 'Sources. . .,' and W. A. Wallace, 'Galileo and Reasoning *Ex Suppositione*. . .,' both cited in note 14 above.

[84] L. M. De Rijk, 'The Development of *Suppositio naturalis* in Medieval Logic', *Vivarium* 11 (1973), 43–79, especially p. 54. Other expositions of Ockham's theory of demonstration are E. A. Moody, *The Logic of William of Ockham*, Sheed and Ward, New York, 1935, and Damascene Webering, *Theory of Demonstration According to William Ockham*, Franciscan Institute Publications, St. Bonaventure, N.Y., 1953.

[85] *In metaphysicen Aristotelis quaestiones*. . ., Lib. 2, q. 1, Parisiis: Venundantur Badio, 1518, fol. 9r; *Quaestiones super decem libros ethicorum Aristotelis*, Lib. 6, q. 6, Parisiis: Venundantur Ponceto le Preux, 1513, fols. 121v–123r. For a discussion of the first of these texts, and a critique of Ernest Moody's reading of it, see W. A. Wallace, 'Buridan, Ockham, Aquinas: Science in the Middle Ages,' *The Thomist* 40 (1976), pp. 475–483.

[86] For the precise texts see *Opere* 5:357.22, 7:462.18, 8:197.9, 8:273.30, 17:90.74, and 18:12.52. The first of these uses the Italian equivalent (*supposizioni naturali*) but the remainder employ the Latin *ex suppositione* even when the surrounding text is in Italian.

[87] Respondeo: scientiam subalternatam tamquam imperfectam non habere perfectas demonstrationes, cum prima principia supponat in superiori probata, ideoque gignat scientiam ex suppositione et secundum quid. . . – MS Gal 27, fol. 20v. This reading is quoted from the transcription of this manuscript kindly made available to me by Adriano Carugo; for a brief preliminary analysis of the place of this work in Galileo's thought, see Crombie, 'Sources. . .,' (note 14 above), pp. 171–174.

[88] Theorica planetarum non est scientia, nam scientia est effectus demonstrationis, sed in illa nulla invenitur demonstratio, ergo. Est igitur scientia vel secundum quid ratione suppositionis vel opinio tantum, non quidem ratione subsequentis sed ratione antecedentis. – Österreichische Nationalbibliothek, Vienna, Cod. Vindobon. 10509, fol. 198r.

[89] This theme is developed in full detail in another paper in this volume, Peter Machamer, 'Galileo and the Causes,' pp. 161–180.

[90] For a good account of the Aristotelian revival in the late sixteenth century, which locates the work of the Collegio Romano in the larger context of European universities generally, see Charles B. Schmitt, 'Philosophy and Science in Sixteenth-Century Universities: Some Preliminary Comments', in *The Cultural Context of Medieval Learning* (ed. by J. E. Murdoch and E. D. Sylla), Boston Studies in the Philosophy of Science, Vol. XXVI, Reidel, Dordrecht and Boston, 1975, pp. 485–537.

[91] Thus some of the claims made by Heiko A. Oberman, 'Reformation and Revolution: Copernicus's Discovery in an Era of Change', in *The Cultural Context*. . .(note 90), pp. 397–435, and by E. A. Moody in his *Studies in Medieval Philosophy, Science and Logic*, University of California Press, Berkeley, 1975, pp. 287–304 and *passim*, with regard to the role of nominalism in the Scientific Revolution, would seem to require revision in light of the findings reported in this paper.

[92] See John E. Murdoch, 'The Development of a Critical Temper: New Approaches and

Modes of Analysis in Fourteenth-Century Philosophy, Science, and Theology,' being readied for publication; also his 'From Social into Intellectual Factors: An Aspect of the Unitary Character of Late Medieval Learning,' in *The Cultural Context*. . .(note 90), pp. 271–384, and 'Philosophy and the Enterprise of Science in the Later Middle Ages,' in *The Interaction Between Science and Philosophy* (ed. by Y. Elkana), Humanities Press, Atlantic Highlands, N.J., 1974, pp. 51–74.

WILLIAM R. SHEA

DESCARTES AS CRITIC OF GALILEO*

Descartes was born in 1596 a full generation later than Galileo and the two men never met. Galileo was seventy-three when Descartes' first book appeared in 1637 and nowhere in his correspondence does he betray any awareness of the younger Frenchman's existence even though Mersenne sent him a copy of the *Discourse de la méthode*. Descartes heard of Galileo, of course, for Galileo's telescopic discoveries of 1610 created a sensation throughout Europe and were even celebrated in a public lecture at the College of La Flèche when Descartes was a student there. Descartes knew Italian, which was taught to their pupils by the Jesuits, but he does not seem to have read Galileo's Italian works on hydrostatics, the sunspots and the comets that appeared between 1612 and 1623. Between 1623 and 1625, Descartes made an extended trip throughout Italy but he did not call on Galileo who at the time enjoyed the enviable possible of Mathematician and Philosopher to the Granduke of Tuscany. During that period Descartes was wrestling with problems of mathematics and optics and was only marginally interested in the astronomical phenomena that confronted Galileo.

Descartes heard of Galileo's condemnation at the end of 1633 or the beginning of 1634 but even then he made no effort to secure a copy of the *Dialogue on the Two Chief World Systems*. His only reaction was to pledge himself to continued discretion about his own Copernican views. As he intimated to his friend, Fr. Marin Mersenne, he had been vindicated in his decision to leave France for the Netherlands where he could be faithful to his motto "Bene vixit, bene qui latuit".[1] Mersenne sought to arouse his interest by mentioning the experiments that Galileo described in the Second Day of the *Dialogue* as evidence for the daily rotation of the earth, but Descartes was unimpressed. He replied: "I deny them all, although I do not on that account believe that the notion of the earth is less probable".[2] It was not until his friend Isaac Beeckman visited him in August 1634 that he actually saw a copy of the *Dialogue*, by then notorious throughout Christendom. Since Beeckman arrived on a Saturday evening and left on the following

R. E. Butts and J. C. Pitt (eds.), New Perspectives on Galileo, 139–159.

Monday morning taking the book back with him to Dort, Descartes had little time to acquaint himself with the content. But this was enough for him to write to Mersenne, on the day Beeckman left, that Galileo discussed motion "reasonably well" but that he was clearly better when he did not tread the beaten path but went his own way. "Except", he added, "when he talks about the tides, which I find that he rather pulls by the hair".[3] Every natural philosopher in the seventeenth century felt it incumbent upon himself to explain the tides and Descartes had in *Le Monde* used his theory of vortex motion to account for the ebb and flow of the sea. He wanted Mersenne to know that he had solved this problem, but he appeared even more anxious that the Mimim Friar should realize that Galileo stated two laws that he had already discovered:

I am, however, willing to admit that I found in his book some of my own ideas, among them two that I believe I wrote to you about some time ago. The first is that the spaces heavy bodies traverse when they fall are to each other as the squares of the times of fall, that is to say that if a ball takes three moments to fall from A to B, it will only take one to continue from B to C, etc. I expressed this with several qualifications for, as a matter of fact, it is never as entirely true as he thinks he demonstrates.[4]

Hence it would seem that the law of free fall, which states that the space traversed is proportional to the square of the time during which the body falls was discovered independently by Descartes before its publication by Galileo. This claim deserves to be investigated.

Descartes first attempt to formulate the law of freely falling bodies goes back to 1618 when he attended the Military Academy founded by the Prince of Nassau in Breda. It is in this city that he first became acquainted with Isaac Beeckman, a Dutch scholar who had been investigating the acceleration of falling bodies. Beeckman was impressed with Descartes' mathematical gifts and he asked him whether the distance a body falls could be determined on the following hypothesis which he had framed:

In the first moment, the body traverses as much space as is possible given the attraction of the earth. In the second moment, it perseveres in this motion to which is added a new attraction so that twice the space is covered in the second moment. In the third moment, the double space perseveres to which is added a third by the attraction of the earth, so that in this moment three times the first distance is traversed.[5]

In the light of this hypothesis, Beeckman asked specifically: "How far will a body fall in one hour if we know how far it falls in two hours?"[6] The question is straightforward but it is formulated in a way that is unusual to the modern reader who would find it more natural to raise the converse question: "How far will a body fall in *two* hours if we know how far it falls in *one* hour?" But Beeckman viewed motion as a process from an initial to a terminal point as Aristotle had taught generations to do. He visualized a body falling from the top to the bottom of a tower and viewed the total distance as the measurable quantity from which must begin the investigation of the position of the body at earlier instants of time.

Descartes coped with the problem with the aid of a diagram in which he represented the distance by a vertical line and the increase in speed by horizontal lines (see Figure 1). The first 'point or minimum of motion' is

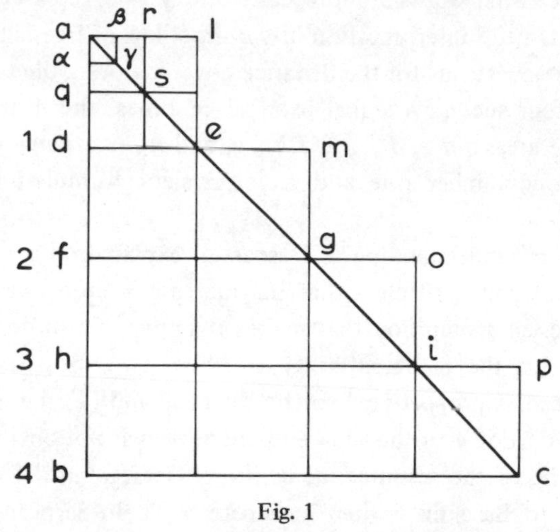

Fig. 1

represented by the square *a d e l*, the second by double that quantity, namely the square *d f g m*, the third by *f h i o*, and so on. But as these intervals are indivisible they must be represented as acting over smaller intervals and Descartes shows how the desired result can be approximated. Divide *a d* in half at *q*. Now the 'minimum of motion' becomes the square *a q s r*, which can in turn be divided along *a q at* α with the new square *a* $\alpha\ \gamma\ \beta$ standing for the 'minimum of motion'. This process is repeated until, at the limit, the parts of the square protruding above the line *a c* vanish and the area above the line is seen to be irrelevant to the computation. The motion of the body accelerating from *a* to *f* and then continuing on from *f* to *b* may be expressed as the ratio of the triangles *a f g* and *a b c*. But the vertical line *a b* stands for the distance, hence the area will stand for the accumulated speeds, and since (by simple Euclidean geometry) the area *f b c g* is three times the area *a f g*, *a b* will be traversed three times as fast as *a f*.[7]

However ingenious, Descartes' solution is wrong since it would entail that bodies fall faster than they actually do. Knowing that the correct relation (first stated by Galileo in the *Dialogue*) is $s = ½ g t^2$, we can see that if x is the distance through which a body falls in time t_1, the time t_2, required to fall through a distance $2x$ is $t_2 = t_1 \sqrt{2}$ and not $t_2 = 4/3\ t_1$ as Descartes' law would have it.

Beeckman, however, misunderstood Descartes' solution and assumed that the vertical line (what we would now called the y-axis) represented the increment of time. On this interpretation, the correct law of free fall follows since the area *f b c g* now stands for the distance covered during the second interval of time. For four successive equal interval of times, the distanced traverse increases as the areas *a d e*, *d f g e*, *f h i g*, and *h b c i*, namely as 1: 3: 5: 7 which is the odd-number rule and an equivalent formulation of the law $s \propto t^2$.[8]

Beeckman's misunderstanding of Descartes' explanation is the more striking since Descartes made it clear that the increase in speed was proportional to the distance, an assumption that makes it impossible to derive the times-squared rule. But the incompatibility of these two propositions: (1) the increase in speed is proportional to the distance and (2) the increase in the distance is proportional to the square of the time was not obvious to any one at the beginning of the seventeenth century. Galileo himself made an error that is similar to Beeckman when he wrote to Paolo Sarpi in 1604 that he could derive the times-squared law from the assumption that speeds in fall increase as the distances from rest.[9] The incompatibility of these assertions only became apparent to Galileo when he worked out the implications of both. The fascinating and complex history of his evolution has been recently explored by Winifred Wison in her important monograph *The New Science of Motion* (*Archive for History of Exacts Sciences* **13** (1974), 103–306).

But the point I wish to make here is simply that Descartes had not reached the correct formulation of the law of free fall in 1618. He was, in some way, hindered from applying Beeckman's potentially fruitful insight that the motion caused by the force of attraction endured. Beeckman had already stated the principle of conservation of motion: a body in motion will persevere in its state unless some new force is brought to bear on it. Descartes later clarified this principle and Newton enshrined it in his *Principia Mathematica* as the First Law of Motion or the Principle of Inertia.[10] But all this lay in the future. In 1618, Descartes still operated within the tradition of medieval physics and he interpreted Beeckman's remarks in the light of the impetus theory. In other words, he substituted the principle of the conservation of force for the principle of the conservation of motion and assumed, along with Buridan, Oresme and the Mertonians, that the speed (and not the acceleration, as we say after Newton) is proportional to the force exerted. Hence a constant force, in this instance the attraction of the earth postulated by Beeckman, produces a constant speed. A body accelerates under the attractive force of the earth because it is more strongly pulled at the end of its downward motion than at the beginning.

Descartes does not seem to have returned to this question until 1629 when Mersenne wrote to ask whether he could determine the length of a pendulum whose period is equal to one half the period of a pendulum of known length. Descartes' reply clearly establishes that he stuck by his erroneous rule of 1618:

> I made the following computation long ago: if a string is one foot long and the weight takes one moment to move from C to B, when the string is 2 feet, it will require 4/3 of a moment; if it is 4 feet, 16/9 of a moment; if 8 feet, 64/27; if 16 feet, $\frac{256}{81}$, which is not much more than 3 moments, and so on for the others.[11]

This passage is instructive because it reveals that Descartes not only retained his earlier law of free fall but was confused about its application to the pendulum. He measured the fall of the bob along the arc CB (see Figure 2) and it was only having receiving a request for clarification from Mersenne that he corrected himself, without explicitly acknowledging his former mistake, and asserted that the distance of fall is not the arc CB but the height KH (see Figure 3).[12]

In this letter to Mersenne, Descartes shows himself to be still operating within the tradition of the impetus theory:

> I suppose that the motion that is once impressed on some body remains in it perpetually if it is not removed by some other cause, that is to say, what once begins to move in a vacuum will move forever with a uniform speed. Suppose therefore that a body located

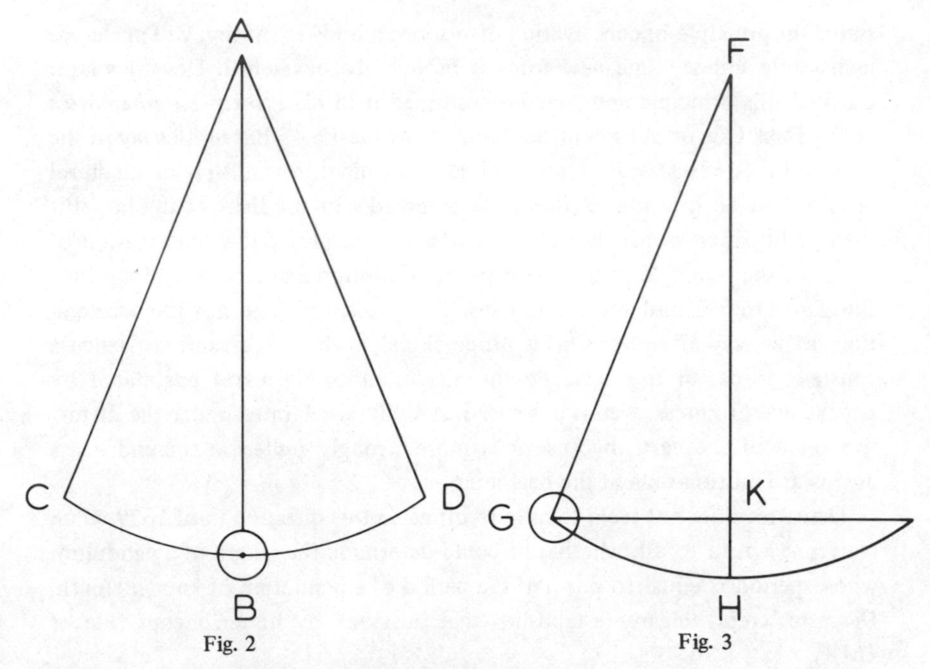

Fig. 2 Fig. 3

at A (see Figure 4) is impelled by its gravity (*a sua gravitate*) toward C. I say that if its gravity left it at the moment its begins to move, it would nonetheless continue in its motion until it reached C, moving at the same rate from A to B as from B to C. This is not the case, however, for its has gravity which presses it downwards and in successive moments adds new forces towards descent. Hence the path, B C, is covered more quickly than AB since the body retains all the impetus with which it moved through the space, A B, and constantly acquires new impetus because of the gravity acting at each successive moment.[13]

It is important to note that Descartes is no longer using Beeckman's notion of attraction. The gravity (namely the weight) is an intrinsic property that gives falling bodies a new impetus downwards at each instant. Acceleration is caused by these '*impeti*' and is therefore measured by their addition. The principle of the conservation of motion is clearly affirmed but it is related to the conservation of these '*impeti*' or impressed forces. Descartes' diagram illustrates how the velocity increased. The first vertical line "represents the force of the speed impressed in the first moment, the second line the force impressed in the second moment, the third in the third, and so on". The triangle ACD thus represents "the increase of the speed of motion during the

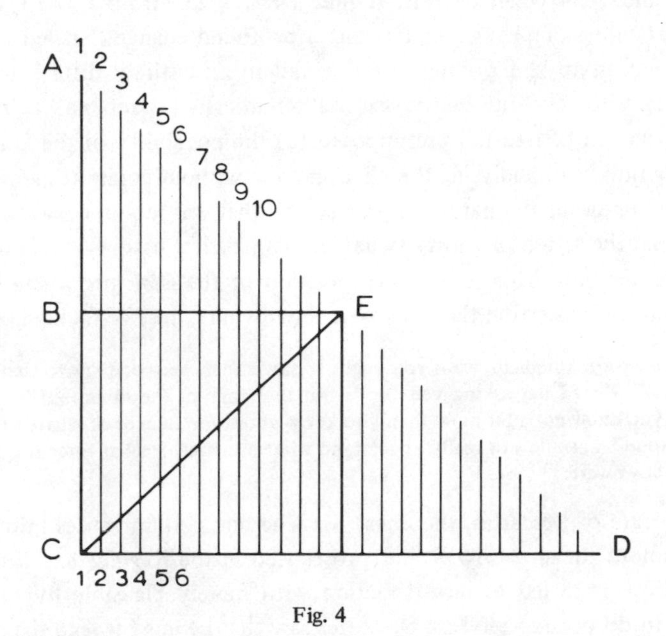

Fig. 4

descent of the body from A to C", the smaller triangle ABE on the upper half of the triangle represents the increase during the first half of the fall, and the trapezoid BCDE the increase during the second half. Since the trapezoid is three times as large as the triangle,

> it follows that the weight will traverse BC three times as fast as AB: i.e., if it descended from A to B three moments, it will go from B to C in one moment, namely it will cover in four moments twice the distance it covered in three moments . . .[14]

That this again is not the times-squared law can be clearly seen for Descartes' law would have bodies fall faster than they, in fact, do. The distances traversed in 3 and 4 moments of time are, according to the times-squared rule, as 3^2 to 4^2, or 9 to 16, not 9 to 18 as Descartes would have it. Now five years later, in 1634, in the letter to Mersenne quoted above, Descartes wrote that he had encountered in Galileo's *Dialogue* ideas already communicated to Mersenne, such as that: "the spaces heavy bodies traverse when they fall are to each other as the squares of the times of fall. . ."

How can we account for Descartes' lapse of memory? Part of the answer lies in the conceptual revolution that took place in Descartes' mind between

1629 and 1634 when he read Galileo's *Dialogue*. Around 1630, Descartes' understanding of physics underwent a profound change that led him to see the solution to the problem of free fall in an entirely different light and eventually to conclude that it was mathematically intractable.[15] Up to 1629, his discussion of free fall presupposed (a) the possibility of the void and (b) the legitimacy of analysing the phenomenon without committing oneself to a determination of the nature of gravity. All that was required was the assumption that the action of gravity (whatever its physical nature) was constant and always identical. This is Galileo's position in the *Dialogue* where he replies to Simplicio's assertion that gravity is the obvious cause of falling bodies:

> You are wrong, Simplicio; what you ought to say is that every one knows that it is called "gravity": What I am asking you for is not the name of the thing, but its essence, of which you know not a bit more than you know about the essence of whatever moves the stars around. . .we do not really understand what principle or what force it is that moves stones downward.[16]

In the case of Descartes, the quest for true and certain causes initiated with the famous dream of 1619 and prosecuted in the *Regulae and directionem ingenii* of 1628 led to dissatisfaction with merely plausible hypotheses. In order to do or redo physics, Descartes saw that he must forego the piecemeal approach of thinkers such as Mersenne and Galileo and proceed, in his celebrated phrase, "selon l'ordre des raisons". He found his model in mathematics, and in the *Discourse on Method*, published in 1637, he summarized the lessons he had learned under the four rules of evidence, division, order and exhaustion. Nothing was to be assented to, unless evidently known to be true; every subject matter was to be divided into the smallest possible parts, each to be dealt with separately; each part was to be considered in the right order, starting by the simplest; and no part was to be omitted in reviewing the whole.[17]

The decision to proceed in an orderly fashion, to begin "with the simplest and clearest ideas" meant, for Descartes, extending mathematical rigour to the entire realm of knowledge and developing, in physics, a system that eschewed notions that could not be defined clearly or were incompatible with the simple cause, contact action, that governed all physical change. *Le Monde* was the embodiment of this new approach which enabled Descartes to give the first clear analysis of the principle of inertia and to deduce its consequence for the motion of bodies in general. In this work, he rejected the notion of a void as unscientific since it entailed the obscure and magical notion of action at a distance, and he offered a mechanical explanation of gravity. Weight is neither an intrinsic property of matter as the Aristotelians held nor the result of the attraction of the earth as Beeckman had maintained, but the consequence of the pressure exerted by the subtle matter that whirls around the earth. The mechanism of free fall is entirely accounted for by contact action: there is no occult pushing or pulling.

The intellectual departure that enabled Descartes to formulate the principle of inertia led, in a sense, to a loss of what he had initially achieved in his treatment of free fall. With the void went the laws that presupposed its possibility. In 1631, Descartes explained to Mersenne:

> I do not disown what I said concerning the speed of bodies falling in the void: for supposing the void, as everyone imagines, the rest can be demonstrated, but I believe that we cannot suppose the void without error.

Nonetheless, Descartes was confident that he could work out the correct law: "But I think that I could now determine the rate of increase in the speed of a stone that falls not *in vacuo*, but *in hoc vero aere*".[18]

Six years later, however, he had not yet determined that ratio and he excused himself to Mersenne by saying:

> it is something that depends on so many others that I could not give an adequate account in a letter. All I can say is that neither Galileo nor anyone else can determine anything concerning this that is clear and demonstrative if he does not first know what weight is and does not have the true principles of physics.[19]

But Descartes knows what weight is. An entire chapter of *Le Monde* is given over to explaining it on the analogy of whirlpools in rivers where solid objects floating in the water are pushed towards the centre. According to Descartes, celestial matter turns faster than terrestrial matter, so that if some object is released above the surface of the earth it is pushed downwards by the celestial matter which has a greater centrifugal force. To Mersenne, who had not been allowed to see *Le Monde*, Descartes vouchsafed the following illustration in 1639.

> To understand how the subtle matter, revolving around the earth, drives heavy bodies towards the centre, fill some round vessel with small lead shots, and mix among this lead some pieces of wood, or of some other matter lighter than lead, that are larger than the shot. Then, making the vessel turn rapidly, you will find that the shot will drive all these pieces of wood or other such material towards the centre of the vessel, just as the subtle matter drives the terrestrial bodies.[20]

But Descartes came to realize that a mathematical formula for the law of free fall would forever elude him. By 1640, he was writing to Mersenne:

I cannot determine the speed with which each heavy body descends at the beginning for this is simply a factual question and depends on the speed of the subtle matter ...[21]

Descartes could not determine the velocity of the subtle matter in the innumerable whirlpools that make up the universe and in his *Principia Philosophiae*, the revised and much enlarged version of *Le Monde* that he published in 1644, not a word is breathed of the mathematical law of free fall ...

If Descartes never had Galileo's *Dialogue* for more than a couple of days, he personally acquired a copy of the *Two New Sciences* as soon as it became available in the summer of 1638. This does not evince a change of heart towards the Italian scientist. Descartes was merely responding to Mersenne's questions about various points raised by Galileo. On June 29th, 1638, he wrote to Mersenne, in what was by now his habitual condescending tone of voice: "As soon as the book is on sale, I'll see it, but only in order to be able to send you my copy, annotated if the book is worth it".[22]

By August 23rd, Descartes had decided that it was not worth it:

I also have the book by Galileo and I spent two hours leafing through it, but I find so little to fill the margins that I believe I can put all my comments in a very small letter. Hence it is not worth that I should send you the book.[23]

The promised letter was written in October. The opening paragraph sets the tone for the ensuing observations:

I find that in general he philosophizes much better than the usual lot for he leaves as much as possible the errors of the School and strives to examine physical matters with mathematical reasons. In this I am completely in agreement with him and I hold that there is no other way of finding the truth. But I see a serious deficiency is his constant digressions and his failure to stop and explain a question fully. This shows that he has not examined them in order and that, without considering the first causes of nature, he has merely looked for the causes of some particular effects, and so has built without any foundation.[24]

As I have already observed, this condescending approach was in keeping with Descartes high opinion of his own method. But the desire to put Galileo down is too manifest not to raise the suspicion that there was more behind his disparagement than the desire to castigate a poor methodology. What galled Descartes, as is clear from a passage in the second half of the letter, was the suggestion that he might have taken something from Galileo. He spurned the very thought:

I have never seen him nor had any communication with him and therefore I cannot have borrowed anything from him. Furthermore, I find nothing to envy in his books and hardly anything that I would wish to own as my own. The best is what he has on music, but those who know me will rather believe that he had it from me rather than that I had it from him. For I wrote practically the same thing nineteen years ago, at a time when I had never been in Italy . . .[25]

If we were, however, to conclude that Descartes was merely being his ironical self, we would miss the significance of music for seventeenth century thinkers. Descartes' earliest work was a treatise on music, the *Compendium Musicae*, written in 1618 as a Christmas present for his friend Beeckman. In this extended essay, Descartes sought to introduce mathematical rigour into current musical theory and he suggested improving the scale by increasing the number of notes from seven to nineteen. He was so proud of his achievement that he waxed indignant when he heard rumours that Beeckman was divulging the new method without acknowledging him as its author. He was so bitter about this alleged plagiarism that he accused Beeckman of downright dishonesty and it was only thanks to the Dutchman's patience and forbearance that a breach between the two was avoided. Descartes' innovation greatly increased the complexities of musical instruments and offered no appreciable advantage except perfect mathematical ratios between the intervals of successive notes. When musicians objected to these perfect consonances on the grounds that the difference between the half-terms could not be discriminated Descartes replied that either they were actuated by the spirit of contradiction or were "hard of hearing".[26] In another letter to Mersenne, he supported his case with an appeal to architecture: to complain that the difference between certain intervals cannot be detected by the ear "is just like saying that the proportions that architects determine for columns are useless since they would look just as nice even if they were off by a fraction or so".[27]

Throughout his entire investigation of music, Descartes was guided by the theoretically clear but experimentally unsound principle that "sound is to the sound, as chord is to chord".[28] In the light of this assumption that the length of the chord directly determines the value of the sound, music became a mere matter of working out mathematical ratios.

Descartes never wavered in his belief that he had made an important contribution to music and as late as 1646 he instructed Constantin Huygens in his system "of constructing a perfect musical instrument".[29] One is reminded of Galileo's proprietorial attitude towards all the celestial discoveries of his day.[30]

But Galileo was clearly in advance of Descartes not only in astronomy but in musicology also. The discussion of consonance at the end of the First Day of the *Two New Sciences* clearly states the fundamental relationships that Descartes never fully grasped: "There are three different ways in which the tone of a string may be sharpened, namely by shortening it, by stretching it, and by making it thinner".[31] Descartes was not even arrested by Galileo's great discovery that the period of a pendulum is proportional to the square root of its length. The only specific point he mentions is Galileo's analysis of the role of weight in lowering the tone to which he, unwarrantedly, objects:

> He says that the sound produced by chords of gold are lower than those produced by chords of copper, because gold is heavier, but it is rather because it is softer. He is wrong in claiming that the weight of a body resists speed more than its size.[32]

Descartes is on better ground, however, when he complains that Galileo's instance of stretching a fine foil of gold does not illustrate the rarefaction of matter. When a piece of metal is ironed out it acquires a greater surface but this does not imply that it's structure is changed.

On the whole, Descartes finds Galileo more given to rhetoric than to rigorous demonstration. Commenting on his device for sliding down a rope without chaffing one's hands by passing the rope through a hollow cylinder, he writes:

> There is nothing in this that is not common, but his way of writing in dialogue form, in which he introduces three persons who do nothing else but praise and exalt his discoveries in turn, greatly helps to recommend his wares.[33]

In other words, Galileo is a communications expert, an excellent salesman, but hardly an original thinker! Descartes does not think very highly of Galileo's determination of the parabolic path of a projectile, not because the reasoning is fallacious – "it is easy to conclude that the motion of projected bodies would follow a parabolic path", he writes – but because the underlying assumption concerning motion in a void is wrong.[34] Galileo's geometrical proofs, "which I have not had the patience to read", are such that "no great mathematical gift is required to find them".[35]

Fortunately, Descartes raised more substantial issues. One of them is the determination of the speed of light. Although Galileo had as late as 1623 entertained the notion that light was transmitted instantaneously,[36] he proposed in the *Two New Sciences* an experiment to calculate its velocity. Two observers, each holding a covered lantern, were to place themselves at night at a distance of two or three miles, and as soon as the first uncovered his lantern the second was to uncover his also. If the exposures and occultations did not occur as they do at a short distance, this would indicate that light takes time to travel from one source to the other. Descartes qualified this experiment as "useless" and offered an alternative which he outlined in a private letter in 1634 but never included in his published writings. Descartes' correspondent had suggested an experiment similar to the one Galileo proposed: an observer would move a lantern in front of a mirror placed at a quarter of a mile and the interval between moving the lantern and perceiving its reflection in the mirror would afford a measure of the velocity of light. Descartes replied that there was another experiment, "often performed by thousands of careful observers that showed that there was no lapse of time between the moment light left the luminous object and the moment it entered the eye".[37] This experiment was provided by the eclipse of the moon.

Descartes' correspondent had conjectured that the speed of light was such that it could cover the quarter of a mile to and from the mirror in one pulse beat. Descartes generously proposed to increase this value by 24 times to 1/24 th of a pulse beat for a quarter of a mile or 1/6 th for one mile. Assuming the current values of 50 earth-radii for the distance of the moon and 600 miles for the length of the earth's radius, this would entail that light takes 5000 pulse beats or roughly one hour to travel from the earth to the moon and back again. Now along a line *ABC*, let *A*, *B*, and *C* represent the positions of the sun, the earth, and the moon respectively; and suppose that from the earth at *B* the moon is being eclipsed at *C*. The eclipse must appear at the moment when the light emitted by the sun at *A*, and reflected by the moon at *C*, would have arrived at *B* if it had not been interrupted by the earth. On the assumption that it takes one hour for the light to make the return journey from *B* to *C*, the eclipse should be seen one hour after the light from the sun reaches the earth at *B*. In other words, the eclipse should not be observed from the earth until one hour after the sun has been seen at *A*. But this is false since, when the moon is eclipsed at *C*, the sun is not seen at *A* an hour earlier but at the same moment as the eclipse. "Hence", Descartes declared using the same word he was to apply to Galileo's suggestion, "your experiment is useless".[38]

The issue of the instantaneous or temporal propagation of light is peripheral to Galileo's physics but it plays an important role in Cartesian mechanism where it illustrates the causal efficacy of contact action in a world permeated with subtle matter. Descartes' saw the instantaneous propagation of light as experimental evidence for his theory and he was even prepared to admit that if an interval of time were detected "my entire philosophy would be completely subverted".[39] Since there is no void in the Cartesian universe, light must be transmitted by a medium whose properties are determined by the nature of corporeal substances. According to Descartes, matter consists in being extended, and this extension is the same as that ordinarily predicated of empty space. It follows not only that two equally extended bodies have the same quantity of matter but that the same part of matter cannot have variable extension.[40] In other words, matter is incompressible and inelastic and in such a medium pressure must be transmitted instantaneously.[41]

A further consequence of the identification of matter with extension is the exclusion of the Democritean model of atoms moving in the void which Galileo sought to reinstate to account for the cohesion of material bodies. In the First Day of the *Two New Sciences*, Salviati, Galileo's spokesman states:

> It is obvious to the senses that abhorrence of a vacuum is undoubtedly the reason why two plates cannot be separated, except with great violence . . . I do not see why this should not also be the cause of the coherence of the smaller parts and even the ultimate particles of these materials.[42]

When a metal is smelted in a furnace, the fire particles reduce it to a liquid state by filling the small intervening vacua which normally hold the metal together. Upon cooling, the fire particles depart and the attractive force of the vacua returns thereby explaining why the metal reverts to its solid state.

Galileo sought to drive home his point with a mathematical argument inspired by the famous problem of the two concentric spheres in the Pseudo-Aristotelian *On Mechanics*, where the difficulty lay in explaining how the larger circle, after one revolution, traversed a line equal to that traced out by the smaller one. Galileo solution was to invoke the existence of indivisible atoms and intervening vacau. He asks his reader to consider two concentric hexagons *ABCDEF* and *HIKLMN* (see Figure 5) such that when the larger

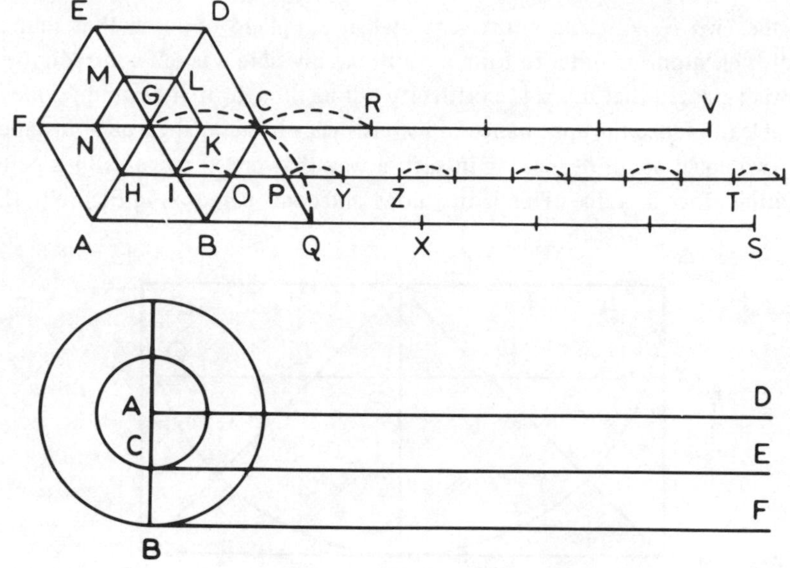

Fig. 5

one moves its side *BC* falls on *BQ* while side *IK* of the smaller hexagon comes to rest on *OP* having skipped over the segment *IO*. At the end of one complete revolution, the smaller hexagon will have touched along *HT* lines equal to its circumference but separated by five 'skipped arcs'. The same reasoning is then applied to circles or polygons of an infinite number of sides. When the smaller circle is rotated, the larger one describes a straight line that is shorter than its circumference because the infinite number of indivisible vacua between the infinite number of indivisible atoms allow for the contraction of the line traced out. If the larger circle is rotated instead, the small circle describes a line greater than its circumference because the atoms and the vacua render expansion possible.

Descartes objected to the demonstration on the grounds that it was mere sophistry:

since the hexagon leaves no void in the space that it traverses, but each of its parts moves with a continuous motion that describes curved lines that fill all the space, we must not consider them, as he does, along a straight line. It does not matter that in his figure the parts IO, PY, etc. of the straight line are not touched by the circumference HIKL for they are touched by other parts of the surface ABC, and hence are not more empty than the parts OP, YZ, etc.[43]

In the *Two New Sciences*, the Aristotelian Simplicio objected that joining indivisible atoms in order to form a continous divisible was self-contradictory. Salviati granted that this was a difficulty but he thought that he could remove or at least reduce the appearance of inconsistency by adducing a clear instance of two equal solids decreasing in such a way that one terminates in a point and the other in a line. This is the 'bowl and cone' paradox. In Figure 6, the

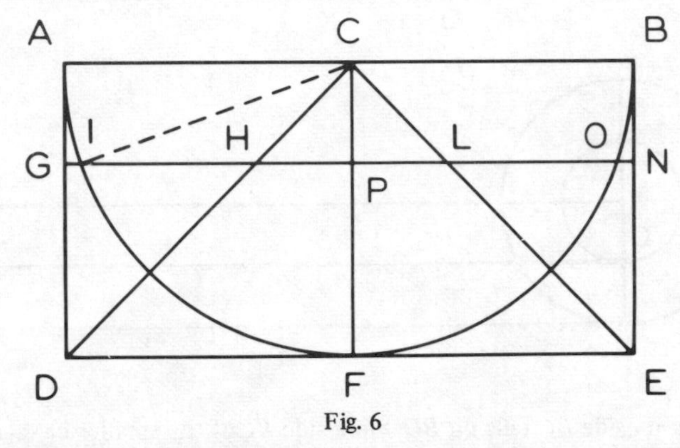

Fig. 6

plane *GN* cuts the cone and the bowl in such a way that part *CHL* of the cone is always equal to the part of the bowl whose profile is represented by the triangles *GAI* and *BON*. As the line *GN* is raised towards *AB*, the areas decrease in the same proportion until in the end of cone becomes a point and the bowl the circumference of a circle. "It appears, therefore, theat we may equate the circumference of a large circle to a single point".[44]

Descartes correctly saw through this specious argument: "*in forma* we can only conclude that a line or a surface is not a greater solid body than the point, not that it is not greater in terms of absolute size".[45] Bonaventura Cavalieri, Galileo's disciple had made much the same objection to his master:

You say that if all the lines of two surfaces are equal when they decrease equally their last diminutions should also be equal. But this is not the case with the bowl and the cone: in the first we are left with a circumference and in the other with a point, which is infinitely smaller than the circumference".[46]

In other words, both Descartes and Cavalieri objected to Galileo's identification of mathematical entities with real physical bodies. That this objection should have come from two of the finest mathematicians of the seventeenth century tells us much about Galileo's mathematical 'Platonism' or his commitment to the belief that "the book of nature is written in mathematical language".[47] However much he may have praised the *method* of mathematics, Descartes never equated understanding physical reality with the knowledge of its geometrical structure. In Rule 14 of the *Regulae and directionem ingenii*, he accuses geometricians of "confusing the clarity of their subject" when after saying that a line has no breadth and surfaces no depth they subsequently

> wish to generate the one out of the other, not noticing that a line, the movement of which is conceived to create a surface, is really a body, and that, on the other hand, a line that has no breadth is merely a mode of the body.[48]

Descartes is often faulted for his uncritical reliance on the mathematical model of deduction. This is true if we consider his methodological programme but it is not relevant to his assessment of the role that mathematics plays in our understanding of nature. Galileo was more committed to mathematical realism and he entertained a broader belief in the explanatory power of mathematical concepts. He never wavered in his faith in mathematics as the source of the notions that enable us to know the real world of nature. In this respect, he was closer to Plato than Descartes was. Galileo felt that the meaning of physics could be completely captured in the language of geometry and he refused to recognize that points that have no breadth or thickness but only position and lines that have length but no width are mere abstractions. He would have seen Descartes' strictures against geometers as evidence that the French philosopher had not freed himself from the shackles of Aristotelianism. When Simplicio, in the *Dialogue on the Two Chief World Systems*, says:

> mathematics may do very well in the abstract, but they do not work when applied to sensible and physical matters. For instance, mathematicians may prove that *sphaera tangit planum in puncto* . . . but in practice things happen otherwise,

he meets with Salviati's scathing reply that only ignorant people believe "that a physical sphere touches a plane at more than one point".[49]

Descartes' dissatisfaction with Galileo's approach was to be voiced once more during 1638 when a third work of Galileo was brought to his notice. This was Mersenne's French version of Galileo's early work on mechanics which had been circulated widely in Italy but had not appeared in print there. Commenting on Galileo's analysis of the lever, he wrote: "he explains very well *quod ita sit*, but not *cur ita sit*".[50] The reason for Descartes' indictment of Galileo for merely stating the *how* without providing the *why* was Galileo's use of velocity and displacement as interchangeable in his consideration of simple machines. In an earlier letter of Mersenne, Descartes had insisted that displacement alone could serve and that reference to velocity was an error the more dangerous as it was more difficult to recognize:

for it is not the difference of velocity which determines that one of these weights must be double the other, *but the difference of displacement* [*mais la différence de l'espace*], as it appears, for instance, from the fact that to raise a weight F by hand to G it is not necessary, if we wish to raise it twice as fast, to employ a force that is exactly double of the one used the first time; but it is necessary to employ a force that is more or less than twice as large according to the varying proportion that this speed can have with the factors of resistance, *whereas to raise it with the same speed twice as high a force that is exactly double is necessary.*[51]

The same ratio holds for velocities and displacements in simple machines and on the strength of this identity Galileo could explain *how* the balance worked. But he failed to see that only the ratio of displacements explains *why* the proportion of force varies as it does. Descartes' insistence on displacement may also have been prompted by his reluctance to get involved in a discussion of velocity, a notion that presupposed a consideration of weight and, hence, in his system, a full account of the action of interlocking whirlpools of matter, a task he knew to be impossible. Nonetheless his distinction between velocity and displacement produced a major step forward in conceptual clarity. Galileo has seized on Archimedes' law of the lever as the means of reducing motion to quantitative treatment and, by means of the principle of virtual velocities, had extended the principle of the lever to all the simple machines. In all instances, the governing principle is the equality of the product of the size times velocity (*mv*) at one end of the lever to that at the other. The *moments* (moment) of the lever thus easily transforms itself into the *moments* (momentum) of the moving body. Since both ends of the lever move in identical times without acceleration, it is immaterial whether one uses the virtual velocities of the two weights or their virtual displacements. Velocities must be in the same proportion as displacements, and Galileo did not realise that the equivalence holds only for the lever and analogous instances in which a mechanical connection ensures that each body moves for the same time, and in which, because of equilibrium, the motion involved is virtual motion, not accelerated motion. The ambiguity built into the model of the lever was to bedevil scientists throughout the century as they sought to interpret real velocities in categories that were valid for virtual velocities. Descartes was not free from the dominance of statics in his characterization of the force of a body's motion as the product of size times velocity but in his treatment of simple machines he clearly saw that the product of size times displacement could be separated from the proportions of the lever and employed as an independent principle. At a later date, a young man, Gottfried Wilhelm Leibniz, was to use it to demonstrate, at Descartes' expense, that the force of a body's motion cannot be proportional to its velocity. But this incident belongs to the history of Leibniz as critic of Descartes.

McGill University

NOTES

* The author wishes to thank the Canada Council for its generous support of his research.

1 Letter to Mersenne, April 1634, in C. Adam and P. Tannery, *Oeuvres de Descartes* 1897–1913. Reprint. Paris: Vrin, I. 286.

2 *Ibid.*, p. 287.

3 Letter to Mersenne, 14 August 1634, A.T., I, 304. Beeckman thought more kindly of Galileo's theory: "... puto eam rationem dignam esse consideratione et meis principiis nullo modo adversantem" (Isaac Beeckman, *Journal 1604–1634* (ed. by C. De Waard), Vol. III, Martinus Nijhoff, The Hague, 1945, p. 171. Descartes outlines his tidal theory in chapter 12 of *Le Monde* (A.T., XI, 80–83) and in the fourth part of his *Principia Philosophiae* (A.T., VIII–1, 232–238). To Mersenne, he confided: "c'est une des choses qui m'a donné le plus de peine à trouver" (letter of November or December 1632, A.T., I, 261).

4 *Ibid.*, pp. 304–305.

5 A.T., X, 58.

6 A.T., X, 60.

7 A.T., X, 75–77.

8 A.T., X, 58–61.

9 Letter to Paolo Sarpi, 16 October 1604, in A. Favaro (ed.), *Opere di Galileo*, Barbera, Florence, 1890–1909, X, 115–116.

10 Beeckman never saw that the principle of rectilinear inertia was incompatible with the assumption that circular motion was inertial. The following entry in Beeckman's diary illustrates the nature of the confusion under which he, as well as Galileo, laboured. "Dictum est mihi hodie qui est dies 11 octob. 1629, Patrem Paulum Servitam Venetum sentire idem quod ego, ut ante saepe patet, de motu, videlicet *quicquam semel monetur, id semper moveri nisi impedimentum accedat*, eoque probasse aeternitatem motus in

coelo a Deo semel motis" (*Journal*, Vol. III, p. 136). The Venetian Servite is Paolo Sarpi, Galileo's friend during his stay in Padua. We see how the necessity of explaining the eternal revolution of the heavenly bodies led, in the absence of a mechanical theory such as Descartes' or a theory of gravitation such as Newton's, to a failure to grasp the implication of the law of inertia.

[11] Letter to Mersenne, 8 October 1629, A.T., I, 27–28.

[12] Letter to Mersenne, 13 November 1629, A.T., I. 73.

[13] *Ibid.*, pp. 71–72. It is interesting that Descartes who had begun his letter in French switched to Latin, the language in which the impetus theory was normally discussed. Latin provided, as it were, a groove along which thought about motion ran only to smoothly.

[14] *Ibid.*, p. 73.

[15] See on this question Ferdinand Alquié, *La découverte métaphysique de l'homme chez Descartes*, Presses Universitaires de France, Paris, 1950.

[16] Galileo Galilei, *Dialogue Concerning the Two Chief World Systems* (transl. by Stillman Drake), Univ. of California Press, Berkeley, 1962, p. 234. In the *Opere*, VII, 260–261.

[17] *Discourse de la méthode*, A.T., VI, 18–19.

[18] Letter to Mersenne, October or November 1631, A.T., I, 228.

[19] Letter to Mersenne, 22 June 1637, A.T., I, 392. See also his letter to Mersenne, 11 October 1638, A.T., II, 385.

[20] Letter to Mersenne, 16 Oct. 1639, A.T., II, 594. The mechanism whereby the subtle matter exerts pressure on falling bodies is described in *Le Monde* (A.T., XI, 72–80) and in the fourth part of the *Principia Philosophiae* (A.T., VIII–1, 212–217).

[21] Letter to Mersenne, 11 March 1640, A.T., III, 36.

[22] Letter to Mersenne, 29 June 1638, A.T., II, 194.

[23] Letter to Mersenne, 23 August 1638, A.T., II, 336.

[24] Letter to Mersenne, October 1638, A.T., II, 380.

[25] *Ibid.*, pp. 388–389.

[26] Letter to Mersenne, 15 May 1634, A.T., I, 295.

[27] Letter to Mersenne, April 1634, A.T., I, 286.

[28] *Compendium Musicae* A.T., X, 97. Descartes also subscribed to the erroneous belief that the speed of sound was determined by its pitch: "Ce que vous dites que le son aigu s'étend plus viste que le grave est vrai en tout sens; car il est plus viste porté par l'air, à cause que son mouvement est plus prompt; et il est plus viste discerné par l'oreille . . . (letter to Mersenne, January 1630, A.T., I, 107).

[29] Letter to Huygens, 1646, A.T., IV, 678–680. In his notebook of 1619–1621 known as the *Cogitationes Privatae*, Descartes already noted instructions for an "instrument de musique fait avec une précision mathématique" (A.T., X, 227).

[30] Galileo, *Opere*, II, 517–518.

[31] Galileo, *Two New Sciences*, p. 100. In *Opere*, VIII, 143.

[32] Letter to Mersenne, 11 October 1638, A.T., II, 385.

[33] *Ibid.*, pp. 381–382.

[34] *Ibid.*, p. 387.

[35] *Ibid.*, p. 388.

[36] Galileo, *Il Saggiatore, Opere*, V1, 352.

[37] Letter to an Unknown Correspondent, 22 August 1634, A.T., I, 308.

[38] *Ibid.*, p. 310. On Galileo's experiment: "Son experience, pour sçavoir si la lumière se

transmet en un instant, est inutile; car les Eclipses de la lune, se rapportant assez exactement au calcul qu'on en fait, le prouvent incomparablement mieux que tout ce qu'on scauroit esprouver sur terre" (A.T., II, 384).

[39] *Ibid*., p. 308. Using Roemer's determination of the speed of light, which gave eleven minutes as the time required for a ray from the sun to reach the earth, Huygens was able to show why the eclipses of the moon did not provide the reliable test that Descartes believed. (See A.I. Sabra, *Theories of Light from Descartes to Newton*, Oldbourne, London, 1967, pp. 203 ff.).

[40] See *Principia Philosophiae*, Second Part, act. 19, A.T., VIII–1, 51.

[41] Sabra notes that Newton himself adopted the doctrine of instantaneous propagation of pressure through an incompressible medium in Bk. II, Prop. XLIII of the Principia (Sabra, *Theories of Light*, p. 56).

[42] Galileo, *Two New Sciences, Opere*, VIII, 66.

[43] Letter to Mersenne, 11 October 1638, A.T., II, 382–383. In a letter to Galileo, 3 March 1635, Antonio de Ville solved the problem of the two concentric spheres much more clearly by printing out that the motion of the small circle is the outcome of two motions in the same direction, namely (a) the rotation of the small circle on itself, and (b) the motion of translation impacted to it by the large circle (*Opere*, XVI, 225–227).

[44] Galileo, *Two New Sciences, Opere*, VIII, 75. In the English transl. of Crew and de Salvio, p. 28.

[45] Letter to Mersenne, 11 October 1638, A.T., II, 383.

[46] Letter to Galileo, 2 October 1634, *Opere*, XVI, 136–137.

[47] Galileo, *Il Saggiatore, Opere*, Vi, 232.

[48] A.T., X, 446.

[49] Galileo, *Dialogue, Opere* VIII, 229–230.

[50] Letter to Mersenne, 15 November 1638, A.T., II, 433. Mersenne's translation, *Les méchaniques de Galilèe* had been published in Paris in 1634.

[51] Letter to Mersenne, 12 September 1638, A.T., II, 354.

PETER MACHAMER

GALILEO AND THE CAUSES*

The literature on Galileo's methodology, or, if you like, his philosophy of science, is replete with reiteration of dichotomous terms, attempting to characterize Galileo's work, necessarily or for the most part, as an instance, of a type. The terms 'Platonism/Aristotelianism,' 'Mathematical/Experimental, 'Rationalist/Empiricist' have been used to describe Galileo's work. What I hope to do in this essay is to provide a way of looking at Galileo which will undercut the force of such dichotomies and which at the same time will be more faithful to the 16th- and early 17th-century traditions of methodological discussion. I hope to make plausible the claim that Galileo *is* in a tradition, but one which has not been sufficiently recognized and has only begun to be studied. The tradition is that of the mixed sciences, which is itself a tradition blending mathematics and physics (or natural philosophy), blending Platonic (or neo-Platonic) and Aristotelian elements, blending reason and observation. It is this tradition I shall argue that Galileo takes on from the late 16th-century thinkers and which can be seen in all his works, even in the much studied *Discorsi*. Indeed, in Section II of this essay I shall concentrate my analysis almost wholly upon the *Discorsi* as published, assuming that if I can make my case plausible for that work the rest of the Galilean corpus will come into line also.[1]

There are a few apologies and caveats before proceeding. I shall not do more than vaguely sketch the historical tradition into which Galileo falls. This is because my present understanding does not extend far enough. Second, though I shall concentrate my efforts primarily on *Discorsi* I shall not attempt to fit the whole of it into my analysis. Also we should be wary of the fact that Galileo, despite his taking of the title 'Philosopher' upon his return to the Medici, is not in many ways a philosopher. I become increasingly more convinced that even though Galileo is working within this tradition of the mixed sciences he is not overly given to philosophical or methodological reflection in any kind of cohesive or systematic fashion. This fact plus his penchant for rhetorical excess means that any attempt to determine Galileo's

R. E. Butts and J. C. Pitt (eds.), New Perspectives on Galileo, 161–180.

methodology or his philosophy of science will be largely reconstructive and will be subject to some apparently contradictory passages from his writings. I shall be satisfied if I can provide a sketch of his *modus operandi* and the character of the concepts involved using the concept of a mixed science, fitting that into a traditional causal framework, that will reconcile some of the apparent tensions and contradictory claims in Galileo's own work and in much of the critical literature on him.

I

Stillman Drake in a footnote to his new translation of *Discorsi* remarks that the rejection of causes is Galileo's most revolutionary proposal in physics, in as much as the traditional goal of science was the determination of causes.[2] Counter to this kind of claim William Wallace has pointed to many instances in Galileo's work where causal language is present and where Galileo explicitly says he is looking for causes. Neither author has analyzed the notions of causality which occur in Galileo's writings.[3]

On the face of it – if we stick to the *Discorsi* since this is where Drake claims Galileo's mature methodology is – Wallace is clearly right. Throughout *Discorsi*, in innumerable places (or at least in too many places to bother to count), Galileo uses the words 'Cagione,' 'Ragione,' 'Causa' (in Italian and Latin) 'Ratio' and cognate expressions.[4] I shall attempt to show that though Galileo does use such causal language with serious intent, there is a sense in which Drake is right about Galileo's unconcern for causes; Galileo is, for the most part but not always, unconcerned about extrinsic, efficient causes. This is one aspect familiar to those who deal with the mixed sciences. Galileo is concerned very much with formal and final causes, and sometimes material causes. This is a standard concern, and was since at least the time subsequent to Aristotle, of someone working with the mixed sciences of optics, harmonics, astronomy, or mechanics.

A mixed science you will recall is one of those sciences which examine mathematical objects *qua* physical (as contrasted with mathematics – at least for Aristotle – which examine physical objects *qua* mathematical). (*Physics* 193b 31 f., *Mechanics* 847a 27). The mixed sciences never assumed pride of place in the curricular strucures of antiquity, though they were always present and became highly cultivated by the artisans in the 16th and early

17th-centuries. My historical thesis is that by the middle of the 16th-century and continuing through into the 17th the mixed sciences became much more significant than ever before in history and provided the general model within which not only Galileo but also Descartes, Hobbes, and many other great men worked.[5]

Though this historical claim needs much support, for present purposes a more general point must start the discussion. The mixed sciences flourished especially well in various neo-Platonic schemes wherein the hierarchies or principles of being received mathematical treatment but did so in completely Aristotelian causal terms. Of importance here is the great revival of systematic neo-Platonism coming during the last part of the 16th century (much of which has been misleadingly called 'hermetical').[6] The names of Telesio, Cardano, Patrizi, as well as earlier thinkers (Roger Bacon, Cusanus) and probably later or contemporaneous men like Kepler, Suarez, Buonomici are relevant here. In all these men Aristotelian causal inquiry is blended with mathematics and with metaphysical and epistemological concerns redolent of classical neo-Platonism.

Given such a general, widespread outlook and concern with natural philosophy and how it fits into such systems, it should not seem surprising that the work of the Greek geometers was treated with a new respect. It is plausible to think that it is in part because of this general framework that the rediscovery of Archimedes was so important (though this is too simple for it was not a one way influence). The combination of the neo-Platonic and geometrical elements were part of what gave rise to the mechanical tradition in the late 16th and early 17th century (the many *De Motu* and *De Mechanica* treatises).

Another influence was operative on Galileo and men in similar positions, and in the same direction. Galileo's training in mathematics with Ricci indicates to us that he was also part of the engineer-artisan tradition of the 16th century. This non-university, more practical tradition likewise was concerned with the mixed sciences and with the explanatory power and epistemological certainty provided by mathematics as applied to natural phenomena.[7] In this regard it is of interest that Galileo not only studied such sciences as perspective (optics) and mechanics, but apparently taught perspective, at one time, to artists.[8]

If we add to this knowledge we have of Galileo's practical mathematical training, the knowledge recently provided by William Wallace, by A. C.

Crombie and Adriano Carugo, and by Paolo Galluzzi concerning Galileo's study of aspects of the Thomistic, Jesuitical and, more generally, the medieval tradition of the natural sciences and demonstration, an interesting picture emerges.[9] It becomes plausible to see Galileo having been trained at first in the artisan tradition trying to make himself more 'respectable' under the influence of Christopher Clavius and others at the Collegio Romano. Galileo seems to have begun, around the year 1590 according to Wallace's figuring, to attempt to turn himself into a philosopher.[10] A philosopher was, of course, something more than the mathematician. The philosopher studied the causes and the art of demonstration. Galileo, I suggest, might best be looked at as trying to legitimize his mathematical pursuits in the eyes of his university colleagues and the intellectual world at large by showing how the results of that tradition could be stated in traditional causal terms. These terms were provided him by the philosophers he studied. These philosophers in the course of their writings or lecturing on the *Posterior Analytics* and *Physics* of Aristotle concerned themselves in large part with the treatment of causes.[11] They are part of the late 16th-century Scholastic tradition and share many characteristics with the neo-Platonic tradition mentioned earlier. I shall have more in detail to say concerning the importance of this later. Suffice it here to say that Galileo apparently thought he had learned his lesson sufficiently for this is the probable motivation for asking for the title 'Chief Philosopher and Mathematician' when he returned to Tuscany in 1610.

All of the influences so briefly described above come together in Galileo. In this light it becomes somewhat clearer why Galileo could consistently say on the one hand that he was an Archimedean and on the other that he was allied himself with Plato and the Pythagoreans. This confluence of traditions allows him to speculate on the nature of Divine Knowledge, to praise mathematics, to admire and learn from the artisans at the Venetian Arsenal, and to present his theories in causal form.

To see a more specific instance of how these general claims are operative consider the relatively unproblematic case of the science of optics. Optics was continually taught as a mixed science since the Hellenistic curricula was established. It was used by some to solve astronomical problems, by others to provide a model for the structure of the universe, by yet others to provide an epistemological basis for knowledge and certainty. In addition, since the 15th-century it had been taught as the science of perspective to artists and artisans of various occupations. In both areas the study of optics (and vision) was a tradition that was alive and thriving in the 16th-century and which continued unabated into the 17th; almost every thinker that historians of philosophy and science study in this period had studied optics extensively. For most, that science has some major role to play in their systems. Light and its study is much more than a metaphor for God and his wisdom in the 16th- and 17th-century.

Optics was considered to be a mixed science as Aristotle said because it treats mathematical lines as physical (as light rays); or as Grosseteste put it, optics shows us physically how to get from points, to lines, to planes (and in Kepler's case, to solid bodies and forces). It is mixed because we have in optics a physical subject matter which is mathematically described. In the traditional, causal terms the formal causes of optics are stated in the language of the propositions and terms of geometry. There is no great problem how mathematics fits into the world. In the more comprehensive systems that used optics as a basic science, light travelling in straight lines is the physical basis by which the story of Genesis is explained. The material causes mentioned in optical sciences are spoken of in two ways. Sometimes the nature of light is considered materially, e.g., as particulate; this is not necessary since most of what one needs to know concerning the nature of light is given by its essential properties such as rectilinear propogation, instantaneous transmission, potential three-dimensional character, etc.; all of which are described geometrically. So that often, in this sense, talk of formal and material causes collapses into talk about natures. The other tradition for treating material causes, in optics and in all mixed mathematical sciences, stems from Aristotle's description of the causes in the *Posterior Analytics* Book II, Chapter 11 (94a 20f.). The material cause is called the necessitating cause and examples of such are said to be the middle terms in mathematical demonstrations. These middles are the material or necessitating cause of the conclusion (of the connecting of the extremes).[12] Optical demonstraitions are, often, geometrical in form; they are mathematical demonstrations. Final causes enter optics in most optical proofs in a variety of forms. Grosseteste's principle of efficiency or least action in nature (as in his refraction law) and a conservation principle assumed in most proofs of the equality of the angles of incidence and reflection are examples.

What is missing in most optical work is the mention of extrinsic efficient causes. The source of light and its mode of transmission is just assumed; it is sufficient for all proofs and for most purposes to which optics was put to formally describe the properties which light has, proving them by appeal to final causes. Such formal and final causes explain the relations between the (formally stated) principles concerned with the nature of light, its transmission and the effects of light (reflection and refraction); these in their turn can be used to explain in the same causal fashion the optical properties of actual instruments, e.g., mirrors and lenses. There is a sense in which *intrinsic* efficient causes are used; these are the *nature* of light itself. Natures, being the intrinsic principles of motion for things, are also said to be formal and material causes (see above). Such a mixed science makes little or no use of *extrinsic* efficient causes, especially when the science in question (as optics was and mechanics will become) is taken as fundamental; when the stuff of that science is treated as the most fundamental natural efficient cause. For optics this fundamental character dates at least back to Aristotle, who in *De Generatione* cites the movement of the sun as the ultimate efficient cause for all sublunar motion (336a 32f.). Many neo-Platonists followed him making light analogously or literally the *forma corporeitatis* and the first moving cause.[13] During the period of our concern the motion of matter becomes the basic efficient cause, and the science of motion (often having optics as a model for its development) becomes the fundamental science for many thinkers. Motion was fundamental for Aristotle, and now it gets mathematicized.

I mentioned, but can not elaborate here the fact that optics provided for many thinkers a model for what a science should be like. There is another mixed science which likewise provided a model.

Mechanics (or not much differently, statics), while not holding the august curricular place of optics, harmonics, or astronomy, is also a mixed science. Mechanics traditionally conceived is the science of weights: the problems of levers, balances, and screws. These are the problems of the pseudo-Aristotelian *Mechanics*, of Archimedes, of Jordanus, and of the 16th-century mechanical treatises. Mechanics is a mixed science in which mathematical concepts are applied without worry to physical objects. The terminology of geometry is used to state the formal causes of the objects considered. Material causes may be given, e.g., to take account of the weight of the lever itself or of the constitution of the body to get, e.g., its specific gravity (so many parts of this kind of stuff to so much of that in this volume) or in terms of the middles of demonstrations. Final causes again have the central place in mechanical proofs, for equilibrium is the standard and equal proportionality or balance is the achievement of rest or final state. Notice again that there is very little mention of extrinsic efficient causes. It is of no concern how the weights got on the balance, little talk of what body actually pushes the lever, no interest in what charge sets the projectile moving. For many thinkers it is probable that they held that there must be an extrinsic efficient cause, but the problems of the mixed sciences are not concerned with that aspect. It is much like the general form of Greek 'projective' geometry. The points, lines and planes of geometry all move to generate their figures, but no one is concerned with what moves them unless this is taken as a way of elucidating their nature (or intrinstic principle of motion). The parallel to the above discussed optics case should be clear.

This mixed science tradition fits well into the tradition born from the 16th-century revival of neo-Platonism. The properties of the mixed sciences noted above can also be seen in most Christian neo-Platonic (or derivative hierarchical) schemes. They are concerned with stating formal, final and material causes the cosmos. They spend little time if any dealing with extrinsic efficient causes. Ultimately the efficient cause of everything is God; this can be assumed and taken for granted. Concern is given to try to explain in the other causal terms *how* God works. Extrinsic efficient causes will later become important in this scheme but only after religious ferment and the con-commitant concerns with heresy bring into prominence a voluntaristic conception of a God with inscrutable purposes. Until the occurs, until Suarez and later Descartes begin to worry about how 'efficient cause' might be used to do the job that 'final cause' once did, extrinsic efficient causes are of much less interest. Certainly Galileo, who is no voluntarist and secure in his faith, does not have Cartesian scruples concerning final causes. Galileo should be seen as belonging to the tradition of explanation fostered by the mixed sciences.

If Galileo is in this tradition we should find him concerned with giving explanations using formal, final and, sometimes, material causes. We should expect little if any reference in most contexts to extrinsic efficient causes.[14]

II

That Galileo uses causal talk I have already affirmed. I must turn to *Discorsi* itself and attempt to lay out some instances of how he uses it. I shall attempt to tie these instances into the general scheme I have just sketched.

On the very first page of *Discorsi* Salviati and Sagredo agree that very often talking with the men at the Venice arsenal helps them in their "investigation of the reason for effects" (11, 49).[15] This remark exhibits the context of the problem for the first day concerning the subject of resistance (*resistenze*) of bodies. Just after this follows the *aporia* or puzzle to start the dialogue. Sagredo notes that in many mechanical devices one cannot reason from the small to the large, i.e., that size makes a difference. Yet, he notes, all reasoning about mechanics has its foundations in geometry where size makes no difference to the geometrical properties. There seems a discrepancy between the principles of geometry and the observed effects in mechanics.

Salviati restates the case and first sets forth, only later to reject, the claim that the discrepancy is due to the imperfection of matter (a claim often put forward by neo-Platonists, e.g., Telesio and some other of the late 16th-century Italian natural philosophers). Salviati points out that matter can be assumed constant (always recalcitrant or whatever), thus it is an eternal and necessary property in all material things. Since mathematics, and only mathematics, describes such eternal and necessary properties, purely mathematical demonstration should be able to explain the apparent discord. The assumption here seems to be that all matter has form, or unchanging nature, and that form or nature can be mathematically described.[16] He claims that it can be demonstrated geometrically that larger devices are always proportionally less resistant than smaller ones.

In this first exchange almost all of our concerns have arisen. Sagredo and Salviati have spoken of finding the causes for effects. They have worried about the relation between mechanical principles which in this context are taken to be generalizations from observed effects and geometrical principles. The contrast is between simple 'inductions' from observations and the true principles underlying such phenomena. They have affirmed that a mathematical resolution to the problem is possible. The resolution is the ascertaining of the cause requisite for allaying the apparent discord. As Salviati affirms three pages later, the recognition of the cause of the effect removes the marvel of it (la riconosciutta cagion dell'effetto leva la maraviglia) (15, 53). Mathematical explanations will allay our doubts. The cause specifically mentioned at this point is a stick's own weight causing it to break when placed horizontally in a wall. The cause here might seem to be an efficient cause. If it is, it is not a traditional moving cause; it is more like those intrinsic causes found in statics and with which Galileo had earlier concerned himself in *De Motu* and *Floating Bodies*.

Salviati next recounts a story to illustrate his claims. He tells of a column laid horizontally to rest on three supports, one at each end and one in the middle. The surprise came when the column cracked right above the middle support. The explanation offered is that one end support had sunk leaving that end of the column in the air. It then broke of its own weight. But Sagredo, in causal terms, questions Salviati's explanation saying he (Sagredo) does not understand the reason why; at the same time he generates another little puzzle as to why thicker nails hold more than double the weight of thinner ones. Ultimately this explanation which seemingly is in terms of efficient causes will be rejected in favor of one using final causes and natures. Sagredo again is the straight man to Salviati's true account.

The Dialogue now really commences in earnest with Salviati telling of the Academician's (Galileo's) new science of resistance. All speculations of the science can be geometrically demonstrated from primary and indubitable principles (15, 54). No persuasion by probable discourse here.

The first move is to ask: what effect (read, 'cause') is at work in breaking a stick or solid whose parts are firmly attached? What is the glue, which like the the filaments of the rope, binds the parts together? Twisting the rope binds the threads. Other bodies whose parts cohere do so by reason of two causes: (1) the celebrated repugnance that nature has against allowing a void to exist, and (2) where the principle of the void is insufficient, some sticky, viscous or gluey substance that will tenaciously connect the particles of which the body is composed (19, 59). Ultimately, as we shall see, (2) is thought *probably* to collapse into (1).

Salviati claims that he can show by clear experiences the nature and extent of the force of the void (on bodies). This sounds very like talk about an extrinsic efficient cause until we examine his example. In a traditional fashion he conceives of two smooth marble slabs which cannot be lifted apart except by a great force. The void, or better the *horror vacui* (*repugnanza al vacuo*), is said to be the cause of this phenomenon. This cannot be a simple (extrinsic) efficient cause. It is, in fact, a final cause specifying an end state which cannot be allowed to occur. That the repugnance to a void serves as a final cause is brought out by Sagredo's immediate response. Sagredo, as he had above, asserts that though experience assures him of the truth of the conclusion (i.e., that the plates do not come apart easily), he is not satisfied about the cause (*causa*) to which the effect is attributed (because, as we shall see, it sounds as though it is an extrinsic efficient cause). To exhibit the source of his dissatisfaction he presents three causal principles which would raise problems for Salviati's causal account, if it were an extrinsic efficient cause that was being stated. The three principles are:

I. A cause ought to precede the effect, if not in time at least in nature; (*la causa debba, se non di tempo, almeno di natura precedere all'effetto*; 21, 60).

II. For a positive effect there ought to be a positive cause; (*che d'un effetto positivo positiva altresi debba esser la causa*; 21, 60).

III. There can be no operation by things that do not exist; (*delle cose che non sono, nissu a puo esse l'operazione*; 21, 60).

Sagredo seems to understand all of these causal principles in terms of extrinsic efficient causes. Only the first (I) seems easily interpretable as a principle affecting the other causes and that only if one takes the 'precedence in nature' reading in an Aristotelian way. The contrast in that case would be the Aristotelian distinction between what is first *for us* and what is first *for nature*. But Sagredo clearly seems to be reading principle I temporally. His puzzle, he says, is that the effect of separating the two surfaces occurs *prior* to the void, which consequently follows the separation. The existence of the void (since it does not exist and is avoided) clearly can not be prior in nature to anything; so Sagredo must have in mind here a temporal puzzle proper only to causal contexts involving independent, external efficient causes. Temporal priority is true only of the relation of such external efficient causes to their effects, not of final, formal or material causes, or internal principles of motion (natures).

Though these three causal principles are not commented upon directly by Salviati, they are apparently rejected or treated as irrelevant, for the discussion shifts. They are irrelevant just because they are principles of efficient causality. Following Sagredo's speech, the Aristotelian Simplicio utters another causal principle, elegant and true, which he claims will solve Sagredo's problem:

IV. Nature does not undertake to do that which cannot be done; (*la natura non intraprende a voler fare che repugna ad esser fatto*; 21, 60).

The existence of a void space, far from being the effieicnt cause of the slabs' cohesion, is 'self-refusing' (*medesimo repugna*). The explanation based on this principle is in terms of final causes; nature prohibits (selects against) actions in consequence of which a void would follow.[17]

Sagredo instead of arguing for his efficient causes, agrees to Simplicio's principle IV and further goes on to ennunciate and draw a consequence from another, accepted causal principle.

V. For one effect there is only one cause; (*di uno effetto una sola e la cagione*; 21, 61).

The consequence drawn claims that if Simplicio is right and if principle (V) holds, refusal of a void should be sufficient to hold together parts of a stone and other solid bodies.

This consequence is played out later by Salviati, where he accepts principle (V) and says "I cannot see why this [repugnance to a void] must not likewise exist and be the cause of coherence between smaller parts, right on down to the minimum ultimate [particles] of the same material." (26, 66). At this point Salviati himself restates causal principle (V) as

V′. For any effect there is one unique and true and most potent cause; (*ed essendo che d'un effetto una sola e la vera e potissma causa*; 26, 66).

'Most potent' (*potissma*) is interesting in this rendering. This is the word for 'power,' 'strength,' etc. but also the superlative form of the word used to translate the medieval Latin 'potentia,' the Greek 'dunamis.' In such causal contexts it means that each effect has as its cause that which has the most potential for it. The cause is potentially the effect; the acorn is potentially the oak tree. The true cause is that which his the greatest potential for the effect. This can be seen as a way of selecting between possible causes. If this reading can be accepted then clearly here again Galileo has in mind final or formal causes: not 'power' in the sense of the power or force of a moving cause, an external, efficient cause. Also the tradition ties talk of unique causes to contexts of final causality, wherein the effect occurring is selected as an unique or most perfect end.[18]

This reading fits textually into what follows in an interesting way. Salviati goes on to consider the possibility of tiny little voids operating on the most minute particles. The voids are said to attract one [particle] against the other (27, 67), and the innumerable little voids multiply resistance (cohesiveness) innumerably. This talk sounds as though the voids were things (reified voids, as it were) and things that have powers of attracting material particles. If meant literally this would be a significant change from the earlier explanation wherein the void appeared only as part of a final cause, as an unactual most despised end.

We are not forced to accept literally this strange shift in talk. Prior to giving the account I have just cited Salviati prefaces his remarks by saying, "I do this not as the true solution but rather as a kind of fantasy full of undigested things that I subject to your higher reflection." Salviati is here telling only a probably story, not one associated with the demonstrative certainty of pure mathematics or geometrical reasoning. In such circumstances one can have recourse to imagined reifications and even talk about such 'things' as if they were external, efficient causes. There are other occasions where one might wish to attempt to isolate efficient causes for certain effects (possibly, accidental effects[19]); but here in this case we are in the realm of *fantasia* – the causes are efficient only as imagined *and* we are forced to imagination for we are dealing with the realm of the unobservable. In this realm *epagoge* cannot be used to ascertain formal causes, to arrive at natures. In the epistemological structure appropriate to the micro-world it seems that true causes can have little part.[20]

Compare the use of 'fantasy' here with that in Drake's favorite passage from the Third Day. In that case Sagredo has started discussing a possible cause of the acceleration of the natural motion of heavy bodies, the force (*virtu*) impressed (157f, 201f.). Salviati comments on this speech that it seems not an opportune time to investigate the cause. Philosophers, he says, have produced many and various *opinions*. Such *fantasies* would have to be examined and resolved, with little gain. So, "It suffices our author that we understand him to want us to investigate and demonstrate some attributes [*passioni*] of a motion so accelerated (whatever be the cause of its acceleration) . . . ." (159, 202). I have italicized 'opinion' to accentuate its contrast with knowledge. Opinions are probables; knowledge is from demonstration and demonstrations ultimately must use mathematical predicates. Presumably, as in the Aristotelian context, Galileo is concerned to stress knowledge of the reasoned fact and not just knowledge of the fact. This latter has a role and comes from observation informed by reason. 'Fantasies' occurs in this context in the same way that we saw before, as imagination. External, efficient causes, the causes Sagredo was beginning to speak about, cannot be used by themselves to provide proper demonstrations. The causal tradition says external, efficient causes can only be used to explain existence *simpliciter.* Such efficient causes leave us in the realm of opinion about the *nature* of the effects and the *nature* of the causes which brought them about. For proper demonstrations we need formal, final and material (necessitating) causes, as found in the mathematical tradition. This use of formal causes as mathematical descriptions I shall talk about next, but here it is important to see that after we leave the realm of observation or experience we leave the realm of knowledge and enter into the arena of opinion. Here there can be no perception of the world as mathematically structured.

Returning to the one cause-one effect principle (V and V′) there is yet another strand of it which is brought up later in *Discorsi* and about which is worth talking. Just after the corollary to Proposition VII (which shows that "the maximum projection . . . will be that corresponding to the elevation of half a right angle:" 245, 296), Sagredo once again marvels at the "force of necessary demonstration." He already knows the fact that half a right angle produced the maximum range for shots, but now he understands (on the basis of Proposition VII and its corollary) the cause (*cagione*) whence (*onde*) this phenomenon.[21] This understanding by cause "infinitely surpasses the simple idea obtained from the statements of others, or even from experience many times repeated" (245, 296). This remark by Sagredo reiterates the important distinction I mentioned above between the fact and the reasoned fact. Clearly, again, it brings out how it is that for Galileo the reasoned fact is provided by providing a mathematical demonstration. Salviati glosses Sagredo's remark by pointing out that the knowledge of one single effect acquired through its causes opens the mind to the understanding and certainty of other effects without need of recourse to experiments; (*e la cognazione d'un solo effetto acquiasta per le sue causa ci apre l'intelletto a'ntendere ed assicurarci d'altri effetti senza bisogno di recorrere alle esperienze, como appunto avviene nel presente caso*; 245, 296).

What is said here bears an obvious relation to the epistemological remarks Galileo had Salviati make toward the end of the first day of *Dialogo.* There

Salviati contrasted the intensive knowledge which we humans have with that extensive knowledge which God has. He remarked that both kinds of knowledge were equal as regards objective certainty

"but the manner whereby God knows the infinite propositions of which we understand some few is much more excellent than ours, which proceeds by ratiocination and passes from conclusion to conclusion, whereas His is done at one single thought or intuition. For example, to attain the knowledge of some property of the circle, which has infinitely many properties, we begin from one of the most simple and, taking that for its definition, proceed with argumentation to another, from that to a third, and then to a fourth, and so on. The Divine Wisdom by the simple apprehension of its essence comprehends, without temporal ratiocination, all these infinite properties which are also, in effect, virtually comprised in the definitions of all things."[22]

Here we see clearly that Galileo's God is not the inscrutable, voluntarist God of Descartes and many later 17th-century thinkers.

If we know the reasons why something happens, we can from these same reasons demonstrate what else must happen; draw out further implications. In order to do this though, we must have recourse to additional definitions (or formal causes) and connect these by material (necessitating) causes in a demonstration. The demonstrations, as is traditional, exhibit the necessity for certain things being true or being the case. This necessity is given to our comprehension by our understanding of the causes of them. In respect of knowing necessity we are like unto God, though he knows the necessity behind all things that happen, behind all properties that are true.

The contrast implicit in the passage from *Discorsi* is that between that is acquired by experience and what is acquired through causal reasoning. Experience by itself cannot provide us with necessity. As he repeats in the letter to Liceti (1640), "experience assures me only of the whether, but brings me no light at all about the how." Only demonstrations can provide us with causes, with answers to how and why questions. Above all mathematics, especially geometry, sharpens the mind and disposes it to reason properly. The study of mathematics teaches us "how to find conclusive reasonings and justifications," (133, 175). This is contrasted in Book II of *Discorsi* with logic which only teaches whether reasonings and demonstrations already discovered are conclusive (133, 175).

Mathematics, especially geometry, is taken as being the preferred set of descriptions by which one describes the world. Mathematical descriptions are preferred because they are used to state the formal causes that are true of the objects, events and happenings so described. Demonstrations are in terms of formal, final and material (necessitating) causes; and further, proof procedures for discovery come about as the result of seeing equivalencies between forms, expressed as necessitating middle terms. Such discovery amounts to seeing that the *Rationem* of this is the same as that; seeing that the same *ratio* applies to these seemingly different things or events. In the famous Third Day Galileo finds that spaces run through equal times by a moveable descending from rest maintain among themselves the same *rationem* as do the odd numbers following upon unity (147, 190; italics mine). Again when he tells that a stone falling from rest acquires new increments of speed, he reports that those additions are made by the simplest and most obvious *rationem* (154, 197). He ends that section by introducing his definition of uniformly accelerated motion, saying that it is seen that we shall not depart far from *recta rationem* if we assume that the intention of speed is made close to the extension of time (154, 198). In all these cases the word I wish to remark upon is 'ratio.' 'Ratio' is used to mean the same as 'reason' or 'formal cause,' sometimes 'definition.' Drake translates it throughout this section as 'rule.' This translation, while not wrong as such, I find misleading unless it is read in a special way. 'Rule' in the proper sense should be a rule like that a just or temperate man follows in living his life. In this Aristotelian sense such a rule, or right rule, is related to our modern sense of 'ratio' for the just or temperate man will live his life according to the mean; according to a certain proper proportion or ratio. I submit that it is in this sense that Galileo also uses, in the Italian, the term 'proporzione,' (e.g., at 159, 202). Of course this sense of 'ratio' or 'proportion' very often will coincide with modern mathematical sense but the medieval Aristotelian influence on its intended sense is present at all times. The use of 'intentionem' and 'extensionem' to describe the important qualities of the event in the last mentioned section supports this Aristotelian reading.

Galileo seems to be invoking this sense of 'ratio' and its cognates in his *On Local Motion* and its commentary. He uses formal causes to provide the grounds for understanding phenomena which we are assured happen by experience. These definitions or reasons show us why heavy things fall and why their acceleration increases. These phenomena of course are verified or experienced, though one need not always continue to have further experiences. This is the reading I recommend for the passage at (159; 202–3) that

Drake takes to show Galileo's verificationism:

> and if it shall be found that the events that shall then have been demonstrated are verified in the motion of naturally falling and accelerated bodies, we deem that the definition assumed includes that motion of heavy things and that it is true that their acceleration goes increasing as the time and the duration of motion increases.

This is not the modern positivistic sense of the doctrine of verificationism. Rather one experiences the definition as holding for the class of events under consideration. This sense of 'verification' is common to all Aristotelian based epistemological doctrines that hold that knowledge comes from the senses.

III

I think from these examples from *Discorsi* we begin to get a picture of how Galileo thinks causal talk works and how mathematics and experience comes together. Mathematics is the primary language because demonstration in mathematics presents us with formal causes. Experience is necessary to tell us what mathematical descriptions are to be applied to an observed phenomenon. Experience may provide us with a set of possible mathematical descriptions, some of which (like the ones at the beginning of *Discorsi*) raise puzzles and are seemingly at odds with common sense. These puzzles are resolved by exhibiting the cause for the observed phenomena. The cause, in the puzzle cases, selects the proper description of the phenomenon and in doing so exhibits the necessity by which the conclusion (*this* description of the phenomenon) follows.

I take this to be the point raised by Galileo in *On Local Motion*, where talking about naturally accelerated motion he writes:

> But since nature does employ a certain kind of acceleration for descending heavy things, we decided to look into their properties so that we might be sure that the definition of accelerated motion which we are about to adduce agrees with the essence of naturally accelerated motion. (153, 197).

The step of experience is not discussed at length by Galileo. We are told to see the world as mathematical; the world is said to be written in the language of mathematics. But how does Galileo justify this claim? The epistemology of seeing the world under the guise of mathematics is traditional. Mathematical properties are properties that objects have and they can be experienced. When we move beyond this as with the microvoids, claims are only probable. What is missing is the account of why the mathematical properties are to be considered essential. That Galileo thinks they are is clear. Why? As far as I know no answer is to be found in the Galileo texts. But clearly he thinks there is no problem about applying mathematics to the world.[23]

Only if Galileo were coming out of a tradition where these principles are taken for granted can we make sense of his use of them. This must be, from the texts we have examined, a tradition that makes primary use of final, formal causes. A tradition which makes little use of extrinsic efficient causes. One that has a place for experience, but that does not require experience in all cases. One that identifies formal causes with mathematical properties and that can take the identity of formal properties to constitute sufficient ground for using that property as a middle term in an explanation. This is the tradition of the mixed sciences, of Archimedes. Aristotle and Euclid have come together. Descartes (and in this respect his follower, Huyghens) will set out in more detail the principles lying behind this tradition and attempt to systematize all knowledge into its form. Historically though, it is more intreesting to note that the account I have given of Galileo is stated in the general methodological remarks made by those who patterned themselves on a Galilean model, those made by the members of the Academia del Cimento:

Now where we may not trust ourselves to go farther, we can rely on nothing with greater assurance than the faith of experience, which (like one that having several loose and scattered Gems, endeavours to fix each in its proper Collet) by adapting the Effects to the Causes; and gain the Causes to the Effects If not at first Essay, as Geometry yet at least succeeds so happily, that by frequent trying, and rejecting, she hits the mark . . .

Here then we ought to carry ourselves as Masterworkmen, to discern between Truth and Error, and the utmost perspicacy of Judgment is but requisite, to see well what really is, from what is not; And to be better able to perform this task, doubtless 'tis necessary to have at sometime or other seen Truth unvailed; an advantage they only have who have had some taste of the studies of Geometry.

(RICHARD WALLER, translator,
Essays of Natural Experiments, 1684; originally, 1667.)

University of Pittsburgh

NOTES

* I have benefited enormously from many conversations with many people. Almost

everyone who participated in the workshop helped me in some way, but especially I must single out helpful discussions with Raymond Frederette, Noretta Koertge, Ernan McMullin, Tom Settle, William Wallace and Winifred Wisan (she also provided me with helpful written comments). In addition, I have to thank Ted McGuire and Bernard Goldstein for their help. I would assume that the fact that I am in debt to so many people from the workshop indicates that it was a great success. More specific acknowledgements are given in footnotes throughout this paper.

[1] This is not meant to imply that there were no changes in Galileo's methodology over time. On the contrary, he can be seen to be reworking certain aspects of his thought; on this see the paper in this volume by Wisan. Despite such changes I would argue that Galileo's general pattern and its constitutive concepts remains essentially the same.

[2] Galileo, *Two New Sciences* (transl. and introduced by Stillman Drake), University of Wisconsin, Madison, 1974, p. 154, footnote 12.

[3] Wallace's discussion is in his *Causality and Scientific Explanation*, Vol. I (University of Michigan, Ann Arbor, 1972), pp. 176f. Wallace does discuss some important aspects of Galileo's demonstrations in his article 'Galileo and Reasoning *Ex Suppositione*: The Methodology of Two New Science' in Boston Studies in the Philosophy of Science, Vol. XXXII (Proceedings of the Philosophy of Science Association 1974), (ed. by R. S. Cohen *et al.*), Reidel, Dordrecht and Boston, 1976, p. 79.

[4] There is a cryptic footnote in Santillana's edition of *Dialogo* (*Dialogue on the Great World Systems*, Salusbury translation, University of Chicago Press, Chicago, 1953; p. 112) where in an apparently irrelevant context he cites a distinction between 'reason' and 'cause' asserted to have been made by Bruno in his *De la Causa* and remarks that Galileo was not unaware of this important distinction. The distinction actually seems to be that between 'cause' and 'principle,' which even in the 16th-century was an opaque distinction, though Bruno does draw it. In general, 'principle' – as in 'first principle' – is used to describe a causal or demonstrative principle.

[5] A similar line is argued by J. A. Bennett in his 'Christopher Wren: Astronomy, Architecture, and the Mathematical Sciences' *J. Hist. Astron.* 6 (1974), 149–84. Bennett briefly discusses the tradition in England and argues that Wren is in it.

[6] J. E. McGuire, 'Active Principles and Neo Platonism: Newton and the *Corpus Hermeticum*' in Robert S. Westman and J. E. McGuire, *Hermeticism and the Scientific Revolution*, (William Andrews Clark Memorial Library, Los Angeles, 1977), pp. 95–142.

[7] It is this tradition of the Mathematical sciences that J. A. Bennett, *op. cit.*, pp. 149–52, calls 'Vitruvian' and which he characterizes as holding that mathematics was the only source of certainty in the natural world and that it was useful and relevant to practical pursuits. This characterization, as far as it goes, seems to fit Italian as well as English thinkers of the time.

[8] This aspect of Galileo's career has been studied by Thomas Settle and it is to him that I am indebted for this connection. Galileo's work with Ricci and his knowledge of the tradition is discussed in Settle's paper 'Ostillio Ricci, A Bridge Between Alberti and Galileo', *Actes du XIIe Congres International D'Histoire des Sciences*, Paris, 1968, Vol. IIIB (Paris 1971), pp. 121–26. During the workshop connected with this volume this side of Galileo's heritage and its importance was again pointed out to me by Tom and specifically he mentioned Galileo's teaching of perspective.

[9] The works referred to are William Wallace's paper in this volume, A. C. Crombie's 'Sources of Galileo's Early Natural Philosophy' in M. L. Righini Bonelli and William R.

Shea (eds.), *Reason, Experiment, and Mysticism in the Scientific Revolution*, (Science History Publications, New York, 1975), pp. 157–76, and Paolo Galluzzi, 'Il "platonissmo" del trado cinguecento e la filosofia de Galileo', in Paola Zambelli (ed.), *Ricerche sulla cultura dell'Italia moderna* (Editori Laterza, 1973), pp. 39–79. I owe Winifred Wisan thanks for referring me to this last, excellent article.

[10] See the paper by Wallace in this volume.

[11] Wallace's paper contains a list of philosophers that Galileo has studied. They include many important medieval thinkers, e.g., Aquinas, Cajetan, Duns Scotus, Toletus.

[12] In his very interesting and helpful article, Gallucci, *op. cit.*, points out that Barozzi and Bianci held the view that mathematical demonstration proceeds through formal and material causes. These philosophers he takes to be, at least indirectly, influences upon Galileo.

[13] Example include Plotinus, Porphyry, Grosseteste, Roger Bacon and many of the 16th-century Italian Natural philosophers mentioned above.

[14] Raymond Frederette pointed out to me one seemingly glaring exception to my expectations. In his last version of *De Motu*, Galileo clearly rejects the theory that lightness is the cause of a body's motion upwards. He claims that those who have held this view did so because they were unable to find an external cause by which the bodies moved (*EN* I, 362; Drabkin translation, p. 120; see also Frederette's 'Galileo's *De Motu Antiquiora*' *Physis* **14** (1972), 346f). Galileo is here arguing that bodies move upward not because of an internal cause but because of an external cause, viz. the extruding action of the medium. But it seems that even here Galileo cannot be read as attempting to establish the explanatory necessity of external efficient causes. He clearly makes use of an internal cause for his explanation of natural downward motion. External causes then seem to be reserved for explaining effects that cannot be accounted for by the natures of the things in question or effects that are 'contrary to nature' (Galileo's phrase in the passage cited above). Thus, it may be that the only occasions on which external causes are to be invoked are when accidents (as opposed to natures) are involved.

[15] In order to reduce notes, in this section I shall put the references to Galileo's *Discorsi e Dimonstrazioni Mathematiche intorno a due nuove scienze* into the text, in parentheses following the quotation or attribution. The first page number will be that of the National Edition (*EN*) (ed. by A. Favaro), Vol. VIII (Barbera, Firenze, 1933); the second page number is that of Stillman Drake's translation, Galileo, *Two New Sciences* (University of Wisconsin Press, Madison, 1974).

[16] Galileo previously discussed such problems with matter and its imperfections in his *Dialogo sopra i Due Massimi Sistemi del Mondo* (1632), *EN* 7, 110f and in Day II, 229f. (Santillana's edition of the Salisbury translation, pp. 95f and 216ff.; University of Chicago Press, Chicago, 1953). In both places Galileo seems to be arguing that mathematical descriptions can be applied to material objects just because the material objects have the property of shape. Considerations concerning the corruptibility or imperfections of matter are irrelevant since they will hold for all matter. Given such constancy the natural philosopher is free to use the supposition that the object in question is like a sphere and then to argue that we can neglect certain sorts of accidents or additions of shape for the purposes at hand.

[17] This use of final causes is seen also in the essential argument in *Dialogo* where Salviati argues from: "Nature does not move to where it is impossible to arrive" (*La natura non muove dove e impossible ad arrivare*) to the claim that circular motion is the only natural

motion (*EN, VII,* 56; Santillana edition; *op. cit.*, 37).
[18] See, for example, Duns Scotus, *De Primo Principio*, Book II.
[19] See note 14.
[20] This does leave open the possibility of using some form of analogical argument. William Wallace has shown that Galileo is familiar (in his 'Juvenalia' and his Logic lectures) with some of the works of Cajetan. It would be worthwhile examining Cajetan's theory of analogy in contexts such as these to determine its influence on Galileo.
[21] Interestingly Drake translates 'cagione onde' as 'reason for,' which is satisfactory as long as the phrase is taken in the traditional causal sense of 'reason.'
[22] *EN*, 128–9; Santillana edition, *op. cit.*, 115.
[23] At the very end of the First Day of *Discorsi* after discussing the periodicity of the pendulum and music proportions, Salviati sketches a basically Aristotelian theory of vision and hearing. It is hinted there that proper perception of the world occurs only when the sense organ is impinged upon by an outside stimulus that is orderly. In hearing such would occur when a well ordered set of vibrations causes the ear to resonate in a proper fashion. This is essentially Aristotle's theory that perception occurs when the organ takes on the ratio of the perceived object. This is another aspect of the doctrine of proportions or ratios that is important in this tradition.

JOSEPH C. PITT

GALILEO: CAUSATION AND THE USE OF GEOMETRY*

1. While Galileo is acknowledged to have been a crucial figure in the scientific revolution of the 16th and 17th centuries, not everyone agrees on the nature of his role. Confusion over what counts as a contribution in the context of extensive conceptual change constitutes the major reason for the disagreement.

There is little or no quarrel over Galileo's scientific accomplishments. What *is* subject to debate is the extent to which he fits any of a number of classifications historians, philosophers and scientists have created. Nevertheless, whichever is used, e.g., Aristotelian, neo-platonist, founder of scientific method, the features the classification is designed to highlight seem basically the same: (1) something began or ended or was in the process of being modified at the time Galileo was working, and (2) in his writings as well as in the conduct of his personal affairs, Galileo managed to reflect the various forces in such a way as to be a focus of attention for both his contemporaries and modern scholars. The purpose of this paper, then, is to outline in some detail an account of the development of science which will permit us to understand what was going on and how Galileo was involved.

When speaking of the development of science, I am referring to the process of refining our conceptual framework by increasing its explanatory power. The process is complicated and perhaps the most truculent problem for analysis concerns the criteria for determining whether a change in theory or methodology will constitute an increase or not.

Kuhn speaks to this issue when he describes what he calls "normal science" (Kuhn, 1970). Unfortunately he also confuses the issue by framing his general discussion in terms of paradigms and paradigm change. Since I am convinced that it is impossible to construct an acceptable account of a paradigm which coherently synthesizes Kuhn's various pronouncements, I will concentrate only on what is important for my present purposes. Kuhn's important insight concerning the promise of a theory does not require paradigms at all. After showing this, I then augment that insight using some ideas of Wilfrid Sellars.

R. E. Butts and J. C. Pitt (eds.), New Perspectives on Galileo, 181–195.

Finally, Galileo's use of geometry is discussed as an example of how conceptual refinements can proceed.

2. Kuhn begins his discussion of normal science by noting that "the success of a paradigm . . . is at the start largely a promise of success discoverable in selected and still incomplete examples." (Kuhn, 1970, pp. 23–24.) I am firmly convinced this is Kuhn's most important idea: that the development of science consists largely in an attempt to capitalize on what appear to be promising lines of attack. Thus, Kuhn notes,

> Normal science consists in the actualization of that promise, an actualization achieved by extending the knowledge of those facts that the paradigm displays as particularly revealing, by increasing the extent of the match between those facts and the paradigm's predictions, and by further articulation of the paradigm itself. (*Ibid.*, p. 24.)

The 'actualization' of the promise a paradigm brings is accomplished through two forms of activity: theoretical and fact-gathering. With respect to the latter, there are basically three types of facts for Kuhn. The first "has shown to be particularly revealing of the nature of things" (*ibid.*, p. 25) and he cites the determination of stellar position as an example appropriate to astronomy. The second "can be compared directly with predictions from the paradigm theory." (*Ibid.*, p. 26.) This is nothing more than the acquisition of experimental data. Finally, we have the determination of those facts which "articulate the paradigm theory, resolving some of its residual ambiguities and permitting the solutions of problems to which it has previously only drawn attention." (*Ibid.*, p. 27.) This would be experimental data designed to establish the unique descriptive and explanatory powers of the theory.

There are also three types of theoretical problems. In normal science theoretical work is involved in developing the theory so as to produce useful facts, in extending the theory to new applications, and finally in theoretical problems of theory articulation. This last is most important for it produces substantial changes in the paradigm. Theoretical changes of this kind are implemented under the impetus of attempting to clarify and extend the fundamental features of the theory. Kuhn's own example here is particularly enlightening. Speaking of Newton's work, he says,

> The *Principia*, for example, did not always prove an easy work to apply, partly because it retained some of the clumsiness inevitable in a first venture and partly because so much of its meaning was only implicit in its applications. For many terrestrial applications,

in any case, an apparently unrelated set of Continental techniques seemed vastly more powerful. Therefore, from Euler and Lagrange in the eighteenth century to Hamilton, Jacobi, and Hertz in the nineteenth, many of Europe's most brilliant mathematical physicists repeatedly endeavored to reformulate mechanical theory in an equivalent but logically more satisfying form. They wished, that is, to exhibit the explicit and implicit lessons of the *Principia* and of Continental mechanics in a logically more coherent version, one that would be at once more uniform and less equivocal in its application to the newly elaborated problems of mechanics. (*Ibid.*, p. 33.)

The interesting feature of Kuhn's discussion here is that he runs together two separate things, clarifying a theory with developing a new one. For to recast the logical structure of a theory is to do one thing and to attempt to reconcile the techniques of the *Principia* with those of the Continental seems to be quite another. He does, however, explicitly note that these efforts to recast a theory have important consequences.

Similar reformulations have occurred repeatedly in all of the sciences, but most of them have produced more substantial changes in the paradigm than the reformulations of the *Principia* cited above. Such changes result from the empirical work previously described as aimed at paradigm articulation. Indeed, to classify that sort of work as empirical was arbitrary. More than any other sort of normal research, the problems of paradigm articulation are simultaneously theoretical and experimental. (*Ibid.*, p. 33.)

At this point it seems appropriate to consider briefly the role of a paradigm for the purpose of expelling it from the discussion. Kuhn has just told us that the problems of paradigm articulation are both theoretical and experimental, where experimental results follow directly from the theory you are trying to articulate.[1] Thus, it would appear that, at least in this case, paradigm articulation is really theory articulation and paradigm reformulation becomes theory reformulation. And, it seems that under the strain of cashing out the notion of the *promise* of a paradigm, the *concept* of a paradigm as something over and above a theory evaporates. If that is the case, then let us replace every occurrence of 'paradigm' by 'theory' in Kuhn's initial characterization of normal science and see what, if anything, is lost.

A. Kuhn's account: Normal science consists in the actualization of that promise, an actualization achieved by extending the knowledge of those facts that the paradigm displays as particularly revealing, by increasing the extent of the match between those facts and the paradigm predictions, and by further articulation of the paradigm itself. (*Ibid.*, p. 24.)

B. Revised version: Normal science consists in the actualization of that promise, an actualization achieved by extending the knowledge of those

> facts that the *theory* displays as particularly revealing, by increasing the extent of the match between those facts and the *theory's* predictions, and by further articulation of the *theory* itself.

It not only appears that nothing is lost, but that a great deal of clarity is obtained. For we know what it is to speak of a theory, hence its predictions, the facts it reveals and an extension of the theory. Furthermore, referring back to Kuhn's example of the effort to reconcile Newton's *Principia* with the mechanics of the Continent, we can use this version of normal science as the first step toward a different theory of scientific development. For as we have seen, using Kuhn's own example, normal science is concerned with the *refinement* of theories. This may consist in the extension of a theory to new domains, or in the attempt to reconcile the theory with a different theory. There is no need to look for some broader concept such as a paradigm. For in extending old theories or in reconciling old theories with new or in reconciling two new theories, the final product is a new theory subject to further refinement. Furthermore, it appears that one of the key features of this kind of change is that it is reasoned. That is, the kinds of refinements on which people are at work can all be explained in terms of the promise of the theory itself. Thus, in the case of reconciling the *Principia* with some Continental theories of mechanics, we can defend the effort by arguing for it on the grounds that the alleged universality of Newton's theory is in doubt because of the difficulty in applying certain aspects of it. Thus, to make good on the inherent plausibility of the assumptions of theory we need to reconcile these assumptions with the more useful and accessible techniques of more limited theories of mechanics. That is, the role of theory itself is what supplies the impetus for reformulation and change.

Obviously the key idea here is that it is possible to give reasons for introducing important and epoch making changes in theories. Unfortunately this point has yet to be proved; so far all that has been demonstrated is that, for Kuhn, an account of theory change can be developed without using the concept of a paradigm. But there is also the beginning of a stronger claim, i.e., that the process of *refining theories* is constitutive of doing science. To make the stronger case, however, it is not enough to show that paradigms have no role in characterizing normal science. An analysis is needed of those periods Kuhn refers to as times of science in crisis showing how our theory of the development of science in terms of refinement of theory covers even the extraordinary situation. This requires some prior discussion of 'refinement'.

As a rough approximation let us take the basic unit of thought to be a conceptual framework. That is to say that thinking occurs only within a conceptual framework. (This obviously needs a great deal of elaboration. Questions concerning the relation between language and conceptual frameworks and the learnability of conceptual frameworks eventually need to be addressed; but, in another paper, on another day.) A conceptual framework consists of a set of categories and a set of rules for operating between and within these fundamental ingredients, rules or categories. The rules for operating within categories are, in effect, constitutive of those categories.

For conceptual change to occur entails a refinement of the categories of the framework. This in turn means a change in the rules constitutive of those categories, i.e., either the logic or the rules used for determining membership. The concept of refinement is, then, to be understood in terms of increasing our ability to make the kinds of distinctions necessary for the further articulation of the inherent promise in a theory or idea by permitting the further development of one of those categories. Much of the impetus for my appeal to this analysis stems from some of Sellars' efforts to deal with these same issues; so perhaps, for purposes of clarity, it is best if we now turn there.

3. In 'Philosophy and the Scientific Image of Man' Sellars calls attention to something he calls the refinement of the original image of man. (Sellars, 1963, p. 10) The "Original image of man" is the phrase Sellars uses to characterize the generalized content of the first conceptual framework man developed. The first men (as creatures capable of conceptual thought) had, on this account, an anthropomorphic view of the world. Thus, volcanos were capable of anger, rivers of revenge, we can speak of trees experiencing the same emotional span as we do, etc. The process of refining this conceptual framework essentially amounted to deanthrogomorphizing nature. The success of this procedure produced the manifest image of man which is basically the view of the world we have in terms of the categories of common sense we employ today. According to Sellars, this process of refinement has two parts to it, empirical and categorical, of which the latter is the most important. Consider now the following condensation of Sellars' line of reasoning.

A fundamental question with respect to any conceptual framework is 'of what sort are the basic objects of the framework?' . . . The questions, 'are the basic objects of the framework of physical theory thing-like? and if so, to what extent?' are meaningful ones.

Now to ask, 'what are the basic objects of a (given) framework?' is not to ask for a *list*, but a *classification* . . .

The first point I wish to make is that there is an important sense in which the primary objects of the manifest image are *persons*. Perhaps the best way to make the point is to refer back to the construct which we called the 'original' image of man-in-the-world, and characterize it as a framework in which *all* the 'objects' are persons. From this point of view, the refinement of the 'original' image into the manifest image, is the gradual 'depersonalization' of objects other than persons . . .

. . . When primitive man ceased to think that of what we called trees as persons, the change was more radical than a change in belief; it was a change in category. Now the human mind is not limited in its categories to what it has been able to refine out of the world view of primitive man, any more than the limits of what we can conceive are set by what we can imagine . . . yet, if the human mind can conceive of *new* categories, it can also refine the old. (*Ibid*., pp. 9–10.)

Thus, (1) we can refine old categories and (2) the development of new conceptual frameworks involves changes in categories.

If we turn to history for examples of situations in which a new conceptual framework is being developed, none seems more obvious than the scientific revolution of the 17th century. Here we have an entire world view under attack, that of Aristotle, and the beginnings of our contemporary view of the nature and structure of the universe and man's place in it. One of the key figures involved in the attack on the Aristotelian view was Galileo. And in his *Dialogue on the Two Chief World Systems* he ostensibly mounts a full scale assault on not only Aristotelian physics, but on Aristotelian principles of reasoning and the basic categories of thought.

And yet, this view of what Galileo was up to quickly falls when just a few familiar items are brought to our attention. To begin with, the Copernican theory of the order of the planets was only an astronomical theory, not a full conceptual scheme. So in defending the Copernican theory alone he is not necessarily arguing that we replace one conceptual scheme by another. Secondly, Galileo not only defends the concept of perfect circular motion for heavenly bodies, but he employs that notion in his own mechanics. Finally, he does not attack the Aristotelian distinction, for example, between rectilinear and circular motion. In other words, there are distinctively Aristotelian dimensions in Galileo's work.

So, if Galileo is not actively engaging in attempting to reject the entire Aristotelian conceptual scheme, what *is* he doing and where does that leave us? Briefly put, Galileo is helping his contemporaries refine the conceptual scheme of the day by developing the conceptual wherewithall to attack the category of causation. As I show below, Galileo argues for the elimination of the teleological dimension of causal explanation by urging the adoption of a method for calculating scientific results which has no room for the role of a final cause. It might be argued, then, that Galileo is thereby creating a new category. But, following Sellars' line of reasoning, this is to confuse refining a category with creating a new one. For Galileo is not providing a new classification of physical things. He does not argue that there are no final causes. Rather, he concludes that, using geometry, descriptions of physical events can be achieved in a rigorous fashion without appeal to final causes.

4. Galileo was instrumental in refining, for scienfific purposes, the Aristotelian mode of causal analysis. This was particularly important, for the continuing growth of science at that time required some such means for extending and systematizing the new description of the universe that was emerging. Galileo advocated the use of geometry as the primary vehicle for establishing correct accounts of the physical structure of the world. In so doing he championed a procedure which permitted the continued growth of scientific enquiry within the context of the prevailing conceptual framework, but in a way which not only removed it from the previous explanatory restraints of that framework, but introduced a new explanatory dimension as well. This is, therefore, an example of what constitutes a refinement in a conceptual framework.

By stressing the geometric mode of analysis, Galileo also (probably unknowingly and unwillingly) opens up the long range possibility of using alternative formalisms to describe the structure of the universe and thereby refine further our understanding by permitting increasingly precise classifications of events. This is not to say this was his intention, but Galileo's move is analogous to the nineteenth-century invention of, or the realization that there can be, non-Euclidean geometries. The development of these geometric alternatives is linked in an important fashion to the development of the new physics and its novel descriptions. Likewise, by providing a formal non-Aristotelian descriptive procedure, Galileo opened the door to the development of alternative formal accounts as well, the key move making possible the development of modern science.

The history of physics from Newton through the present day is the story of the continuing effort to develop a mathematical description of the physical world that is consistent with a causal interpretation of that formalism. The result of the continuing struggle to reconcil these two dimensions of a theory has been a series of mutual readjustments in both the formal structure and the interpretation of it. Kuhn pointed to this kind of activity in the account we discussed earlier of nineteenth century effects to refine the methods in Newton's *Principia.*[3] The ultimate success of this process depends not on whether we can find a completely adequate causal interpretation nor does it depend on our ability to construct a more accurate formal structure *per se*. It relies, rather, on the conceptual latitude which permits further refinement.

For the key issue here is not whether such a coherent and complete account is possible; it is, rather, that the search for such a result is what constitutes the scientific enterprise. To continue to do science requires a conceptual framework rich enough to admit the kind of continuing concept-jostling and readjusting that the process of fitting data to theory demands. If it were either the case that every change in science forces a complete change in conceptual framework or that the possibility of inventing new hypotheses and developing new theories is rendered impossible, then science itself would be impossible.

Now one of the most interesting features of Galileo's move to open up the Aristotelian framework to fundamental change is that his conceptual base is a distinction well entrenched in the Aristotelian framework. This is the difference between *scientia quia* and *scientia propter quid* or knowledge of the fact and knowledge of the reasoned fact. Both constitute scientific knowledge, but for Aristotle the ultimate aim of enquiry is to secure the latter, which involves an explanation of why the fact is as it appears. What Galileo does is to deny the priority of *scientia propter quid* and argue for the superiority of *scientia quia* by relying on the descriptive prowess of a mathematical tool: geometry. The grounds for his argument are epistemological and concern what some might consider to be a theological point concerning the relation between man and god.

> . . . human understanding can be taken in two modes, the *intensive* or the *extensive*. *Extensively*, that is, with regard to the multitude of intelligibles, which are infinite, the human understanding is as nothing even if it understands a thousand propositions; for a thousand in relation to infinity is zero. But taking man's understanding *intensively*, in so far as this term denotes understanding some proposition perfectly, I say that the human intellect does understand some of them perfectly, and thus in these it has as much absolute certainty as Nature itself has. Of such are the mathematical sciences

alone; that is, geometry and arithmetic, in which the Divine intellect knows all. But with regard to those few which the human intellect does understand, I believe that its knowledge equals the divine in objective certainty, and there it succeeds in understanding necessity, beyond which there can be no greater sureness. (Galileo, 1632, pp. 103–104.)

This is an elaboration of an early point Galileo makes, in which he sets up the case for the desirability for the kind of knowledge mathematics provides. There he notes that,

For anyone who had experienced just once the perfect understanding of just one single thing, and had truly tasted how knowledge is accomplished, would recognize that of the infinity of other truths he understands nothing. (*Ibid.*, p. 101.)

The emphasis here is on the manner of coming to know and this is a point he returns to at the end of Day One.

... as to the truth of the knowledge which is given by mathematical proofs, this is the same that Divine wisdom recognizes; but I shall concede to you indeed that the way in which God knows the infinite propositions of which we know so few is exceedingly more excellent than ours. Our method proceeds with reasoning by steps from one conclusion to another, while His is one of simple intuition ... the Divine intellect, by a simple apprehension of the circle's essence, knows without time-consuming reasoning all the infinity of its properties. Next, all these properties are in effect virtually included in the definitions of all things; and ultimately, through being finite, are perhaps but one in their essence and in the Divine mind. (*Ibid.*, p. 104.)

In knowing geometric propositions as mathematical conclusions what one knows is the necessity of the order of things and their properties. The necessity here concerns the subject matter of the proofs, not the concept of necessity itself. More importantly, what one knows is the way things are necessarily ordered, for that is the sense of 'properties' here. Galileo rests his case on the reliability of geometry only because he is convinced of the presupposition that the use of geometry can yield certain knowledge because the universe is organized in accordance with geometrical principles. If the universe were not so organized, then whatever else might be forthcoming from the application of geometry, knowledge of true and certain causes would not. And in this Galileo has made a sharp break with the long-standing geometric-mechanical tradition. For in insisting that the mathematics yields true knowledge, Galileo argues that mathematics has a more powerful use than merely to construct interesting models which serve to save the phenomena.

In Day Two, Galileo once again approaches the problem of the relation between geometry and the world. The discussion begins with Salviati trying to convince Simplicio that a stone falling from a ship's mast will strike its deck in the same place irrespective of the motion or rest of the ship.

Without experiment, I am sure that the effects will happen as I tell you, because it must happen that way; and I might add that you yourself also know that it cannot happen otherwise, no matter how you may pretend not to know it – or give that impression. (*Ibid.*, p. 145.)

The argument proceeds on geometric principles using the concept of a ball on a plane. It is designed to lead Simplicio to confess that he *knows* without having recourse to experience. Again the insistence is on the necessity of knowledge produced independently of experiment. The argument proceeds through a long discussion of the motion of projectiles and finally arrives at the point where Simplicio poses the key question concerning the reliability of abstract geometric reasoning for empirical purposes. Salviati responds with an account of how geometry can be used to aid us in discussion of concrete objects as well as those concerned solely with abstract forms.

But I tell you that even in the abstract, an immaterial sphere which is not a perfect sphere can touch an immaterial plane which is not perfectly flat in not one point, but over a part of its surface, so that what happens in the concrete up to this point happens in the same way in the abstract. It would be novel indeed if computations and ratios made in abstract numbers should not thereafter correspond to concrete gold and silver coins and merchandise. Do you know what does happen, Simplicio? Just as the Computer who wants his calculations to deal with sugar, silk and wool must discount the boxes, bales, and other packings, so the mathematical scientist (filosofo geometra), when he wants to recognize in the concrete the effects which he has proved in the abstract, must deduct the material hinderances, and if he is able to do so, I assure you that things are in no less agreement than arithmetical computations. The errors, then, lie not in the abstractness or concreteness, not in geometry or physics, but in a calculator who does not know how to make a true accounting. (*Ibid.*, p. 207.)

That Galileo is arguing here for a correspondence between the domain of the abstract and that of the concrete seems clear. And in doing he brings to the fore once again the presupposition he invoked when speaking about our ability to obtain true and certain knowledge of the universe. Galileo is not explaining geometry in terms of physical relations and physical relation in terms of geometry. Instead, he is arguing that there is a one to one correspondence between them. Failure to demonstrate the correspondence is not a failure in the geometry, but rather in the individual computing the correspondence. It is essential, of course, to start with correct hypotheses, hence the efforts to eliminate erroneous computations. We are reminded in a number of places that were we only but better versed in geometric reasoning, empirical errors would be minimized. In the example given above the blame for error is placed on the shoulders of the computer. This point is emphasized again when Galileo is discussing Gilbert's account of the Lodestone at the end of Day Three.

> What I might have wished for in Gilbert would be a little more of the mathematician, and especially a thorough ground in geometry, a discipline which would have rendered him less rash about accepting as rigorous proof those reasons which he puts forward as *verae causa* for the correct conclusions he himself observed. His reasons, candidly speaking, are not rigorous, and lack that force which must unquestionably by present in those adduced as necessary and eternal scientific conclusions. (*Ibid.*, p. 406.)

When Galileo notes here that Gilbert's *reasons* are not rigorous he must be speaking metaphorically, for the attributes of rigor apply to structures by which one supplies reasons in support of his conclusions. That being the case, the criticism directed against Gilbert amounts to an attack on the presentation of his reasons for his conclusions and alludes to errors arising from lack of rigor in the drawing of conclusions. In this case, the computer, Gilbert, is at fault for not using the most secure manner to reach his conclusions. Thus the conclusions themselves are in doubt. But, we are led to believe, had he reasoned correctly, i.e., geometrically, then had he begun with true insights he would have concluded necessarily what is the case.

Now this line of reasoning is one with which we are all familiar. Give me true statements and a truth preserving method of reasoning and I will produce truths. Furthermore, none of this was essentially new to Galileo's contemporaries. What is of interest here, however, is not the use of geometry, but two other items. First, there is the development of the realism theme concerning a one to one correspondence between the structures of geometry and the world. Once one initiates a line of reasoning about the world using correct initial claims, then the necessity of the claim about the structure of the world which is the conclusion follows by virtue of the priviledged character of geometric principles. Second, despite the attention paid to the presupposed correspondence, Galileo allows that a description of the structure of the universe is not an account of the causal principles governing the motion of the various components of this system. *But* it can be an explanation if exploited from the perspective of realism.

This most important aspect of Galileo's procedures is to be extracted from the manner in which he works geometry into the discussion and then out again. The very first important argument in Day One of the DIALOGUE presents us with a clear example. First, he considers the typical Aristotelian proof for the existence of three dimensions, then we find Galileo presenting an alternative proof using only the principles of geometry. There then follows an appeal for the continued use of geometry. But along with this appeal there is a subtle set of suggestions. Consider Simplicio's remarks at the conclusion to the proof for three dimensions and Sagredo's response.

Simplicio: I shall not say that this argument of yours cannot be conclusive. But I still say, with Aristotle, that in physical (*naturali*) matters one need not always require a mathematical demonstration.

Sagredo: Granted, where there is none to be had; but when there is one at hand, why do you not wish to use it? (*Ibid.*, p. 14.)

Simplicio has no answer to Sagredo's question. But perhaps we can formulate one for him. One reason for not using mathematical, viz: geometrical, arguments, is that it is not necessary to consider metaphysical claims at the same time. And in cases where the mathematics tends to take on a life of its own, the more one is able to detach one's reasoning from a particular semantics, the less one is likely to find interpretations for the results. For the Aristotelians, the conclusion then is that since you may not be able to provide an interpretation for the conclusion of a mathematical demonstration, you may end with no explanation at all.

The interesting thing is that Galileo doesn't seem to be worried about this; for he has already implicitly substituted an alternative metaphysics for the Aristotelian when he announced that the reason why we can understand so much of the world is due to the geometric mode of divine thought. A corollary to this is Galileo's own presumption here that in many cases, in particular those for which mathematical demonstrations are possible, *the explanation is the demonstration*. What one should keep in mind is the important point that for Galileo explanations are possible without appeal to the causal principles traditionally deemed necessary for scientific knowledge as *scientia propter quid*. He succeeds by virtue of appeal to a presumption shared by all, that we can indeed use geometry to obtain empirical information. But Galileo was not the first to rely on this assumption – the history of Archedimean Mechanics attests to its long-standing role. So does the history of astronomy. But, especially in astronomy, Galileo breaks with the tradition of using geometry to generate *only* observational data and claims more. He, unlike Ptolemy and Osiander, is a realist, not an instrumentalist. Mathematical demonstrations can produce not only testable observations but explanations because the world is ordered in accord with the principles of geometry. And this is the key, for he is insisting that we can extend our *explanatory* abilities by the use of mathematics.

5. One of the more common criticisms of the Aristotelian conceptual scheme concerns the tightness of fit of the basic concepts. This is also one of its strengths. But in the very lack of the kind of latitude necessary for conceptual change lies the seed of explanatory stagnation. The introduction of conceptual changes is the prerequisite for scientific progress and that progress can be traced by following the manner in which fundamental categories are refined by virtue of the introduction of new distinctions within that category, distinctions which open up new possibilities for structuring our knowledge. What Galileo did was to revitalize the Aristotelian framework by working out some further distinctions within the category of causation. This is not to say that he thereby eliminated causes from science or the rest of our basic conceptual framework. What happened was that scientific knowledge, appropriately called, was liberated from the need to appeal to all of Aristotle's four causes. In particular, by arguing that use of geometry gives us knowledge of the necessary structure of the world, the need to appeal to a final cause and thereby invoke teleology is abrogated. Scientific enquiry can now proceed without the restraint of having to fit the results into a teleological universe. But this is not to argue that all purpose has been removed from our understanding of the world. This is only to claim that in providing scientific knowledge it is not necessary to explain the phenomena in terms of final causes. The role of final causes remains in other domains, perhaps to explain man's role in the universe or, at this stage in science, biological development.

Effectively this amounts to refining our category of causation as it operates in the production of scientific knowledge. By eliminating final causes from science, we can no longer classify events in those terms. Furthermore, despite the fact that Galileo is personally convinced of the special epistemological properties of Euclidian geometry, what is important is not geometry so much as the establishment of the explanatory sufficiency of mathematical results. For now it's possible to argue, should this mathematical system prove inadequate, that it should be replaced by another, i.e., the calculus, and only when this can be done can the development of modern science as we know it proceed. The consequence of viewing scientific progress in this way is that progress is marked by our ability to make finer distinctions, thereby revealing the world to be the complicated place it is without destroying the more everyday picture we continue to live with despite what science says.

Virginia Polytechnic Institute and
State University

NOTES

* I am indebted to R. E. Butts, Ernan McMullin, Charles Cardwell, and Marc Lowenstein for many helpful comments on earlier versions of this paper.

[1] We might speculate as to the initial motivation for arbitrarily distinguishing theoretical from empirical paradigm articulation. At best it could be seen as an attempt to do justice to the extremely diverse catelogue of activities that constitute the doing of science. But at worst, it appears to be a crude attempt to extend the concept of a paradigm beyond that of a theory. The truth, I suspect, lies somewhere in between.

[2] Professor McMullin has pointed out that it might be objected here that this is a misleading way to put forth Sellars' views. For, beyond the manifest image Sellars postulates the development of a scientific image and, it is alleged, since the development of the scientific image does not proceed in the same way as the manifest did, via refinement, aren't we being unfair to Sellars? I think not. First, Sellars' scientific image will involve a synthesis of the products of the various sciences, which I am arguing they produce using the refinement model in the first place. Secondly, the production of the scientific image requires a new conceptual scheme, one in which the primacy of 'person' is either replaced or augmented. The justification Sellars invokes for the development of both particular sciences and the framework for the scientific image is explanatory coherence (see Sellars, 1964). Thus, the development of the scientific image should proceed for Sellars in terms basically similar to the development of any individual science or individual theory or the manifest image out of the original image. The criterion, increase in explanatory power (using a combination coherence-essentialist theory of explanation), remains the same. What is changed remains the same since it involves a change in the appropriate constituative rules.

[3] See my 'Revolutions in Science and Refinements in the Analysis of Causation' (with M. Tavel) in *Zeitschrift für Allgemeine Wissenschafts-theorie,* 1977.

BIBLIOGRAPHY

Galileo, G., 1632, *Dialogue Concerning the Two Chief World Systems* (transl. by Stillman Drake), University of California Press, Berkeley, 1970.

Kuhn, T., 1970, *The Structure of Scientific Revolutions*, 2nd Edition, Enlarged, University of Chicago Press, Chicago.
Sellars, W., 1963, 'Philosophy and the Scientific Image of Man', in *Science, Perception and Reality*, Routledge and Kegan Paul, London.
Sellars, W., 1964, 'Induction as Vindication', in *Philosophy of Science* 31.

H.E. LE GRAND

GALILEO'S MATTER THEORY

A revival of atomism stimulated by the recovery of Lucretius' *De rerum natura* and Hero's *Pneumatica* occurred in the sixteenth century. In addition to these ancient theories, a number of original corpuscular theories were propounded, though many of these borrowed heavily from the older traditions. Most of the atomists of the early part of the seventeenth century, though rejecting the plenum of the Aristotelian world-view, still embraced the remainder of that system. They made no attempt to replace the substantial forms and real qualities of Aristotle with the concepts that nature is merely inert matter in motion and that all causation occurs through material contact. Moreover, like the ancient atomists, they failed in their attempts to explain plausibly the properties of gross bodies in terms of the characteristics of their constituent particles.[1]

Galileo, as an exponent of the new mechanical view and as a significant figure in the development of science, naturally enough has attracted considerable attention from historians interested in the development of seventeenth-century atomism. Largely because of the famous passage in *The Assayer* in which he discusses the subjectivity of qualities (Galileo, 1964–66, Vol. vi, pp. 347–350; Drake and O'Malley, 1966, pp. 309–312) and also the references to Democritean atomism in the *Floating Bodies* (*Ibid.*, Vol. iv; *e.g.*, pp. 132–33), many scholars have regarded him as a Democritean or Epicurean atomist.[2] Kurd Lasswitz, for instance, in a work still frequently consulted attributes to Galileo the conjoining of the Aristotelian four elements to a Democritean interpretation of sensation (Lasswitz, 1890, Vol. ii, pp. 37–55; see also Löwenheim, 1894). It is a small wonder that A.G. van Melsen, taking his cue from Lasswitz' analysis, dismissed Galileo's matter theory as "relatively unimportant because his concept of smallest particle is too confused." (Melsen, 1952, p. 112) Marie Boas Hall has more correctly adduced the interpretation of Galileo's ideas on atoms and vacua as Heronic. (Hall, 1949, p. 48; Hall, 1952, pp. 432–37)

Interpretations of Galileo's matter theory as derivative and static represent

R. E. Butts and J. C. Pitt (eds.), New Perspectives on Galileo, 197–208.

a misunderstanding and oversimplification of his all too brief remarks. William R. Shea has recently sketched out a persuasive argument that Galileo's ideas were not fixed, but rather evolved from the discussions on floating bodies to the publication in 1638 of the *Two New Sciences* (Shea, 1970). Shea's analysis of the final theory as presented in the *Discourses* is, regrettably, quite brief. Moreover, he portrays Galileo's matter theory as an outgrowth of certain philosophical problems, as is certainly quite correct, but passes over the specific physical problems to which it provided solutions. It is the intent of this essay both to identify the specific physical problems solved by Galileo's 'mathematical atomism' as presented in the *Discourses on Two New Sciences* and to clarify somewhat the labrynthine discussions occupying the First Day of that work by interpreting them as, *inter alia*, a careful presentation of a highly original matter theory.

Galileo implicitly accepted an atomic view of matter as early as 1612, but only as a working hypothesis. (*Ibid.*, pp. 13–15) He soon confronted the inherent difficulties of the ancient atomic views in explaining various physical phenomena associated with gross bodies. In particular, the phenomena of rarefaction and condensation presented a grave challenge to the credibility of both the Heronic and Democritean approaches. If a dense body be composed of contiguous finite particles, how can contraction be explained without in some way violating the principle of the impenetrability of matter? If, on the other hand, one assumes the existence of void spaces between these particles, and expansion to be the result of increasing these spaces either in number or size, how can the unity of the gross body be explained; *i.e.*, what prevents the complete disassociation of its constituent parts?

Galileo pondered such difficulties for a number of years prior to the *Discourses*. In *The Assayer*, for instance, Galileo responded to 'Sarsi's' contention that motion is the cause of heat as follows:

> I should therefore like very much to have seen explained more specifically how it is possible to rarefy bodies without any separation of their parts, and how this affair of rarefaction and condensation proceeds, of which Sarsi seems to speak with great confidence; for to me, it is one of the most recondite and difficult questions of all physics. (Galileo, 1964–66, Vol. vi, p. 331; Drake, 1966, p. 292.)

Indeed, Galileo's concern with the nature of rarefaction and condensation can be dated to as early as 1614 or 1616.[3] There are other interesting references in *The Assayer*. He writes of the cohesion or adherence of two polished slabs of marble and muses that this adherence may be the only cement binding the particles of solid and hard bodies. Air, and of course water, have no such cohesive force, for he adamantly rejects the contention that some objects float because of the adherence of air to their surfaces and argues consistently throughout his writings that water offers no resistance to penetration and therefore, by implication, has no internal cohesion. He is puzzled, however, that "large particles of water sustain themselves . . . upon the leaves of cabbages." (Galileo, 1964–66, Vol. vi, pp. 322–324; Drake and O'Malley, 1966, pp. 282–83) These seemingly unrelated thoughts in 1634 suddenly were brought together under a common heading: mathematical atomism (*ibid.*, Vol. xvi, p. 163). The result is the First Day of the *Discourses*.

Viewed from this perspective, the First Day is not a haphazard discussion of cohesion repeatedly interrupted by digressions on a variety of topics ranging from atoms to infinity. Galileo carefully delineates in this section of the *Discourses* a novel matter theory related to, but distinct from, classical atomism. This theory, moreover, is mathematical in form, befitting his conviction that mathematics is the basis of objective knowledge. He, however explicitly states neither the physical problems which had vexed him nor his matter theory which solved them, so that it is necessary, in his words, to "note how conclusions that are true may seem improbable at first glance, and yet when only some small thing is pointed out, they cast off their concealing cloaks and, thus naked and simple, gladly show off their secrets." (*Ibid.*, Vol. viii, p. 52; Galileo, 1974, p. 14.)

The point of departure for Galileo's presentation is cohesion. The interlocuters move swiftly from a discussion of the strength of materials to the question: "What effect is at work in the breaking of a stick, or of some other solid whose parts are firmly attached together?" (Galileo, 1964–66, Vol. viii, p. 54; Galileo, 1974, p. 16.) Salviati, speaking for Galileo, answers that the parts of solids cohere by reason of one of two causes, "one of which is the celebrated repugnance that nature has against allowing a void to exist." If this be insufficient, he continues, it may require the auxilliary action "of some sticky, viscous, or gluey substance that shall tenaciously connect the particles of which the body is composed." (Galileo, 1964–66, Vol. viii, p. 59; Galileo, 1974, p. 19.) He expounds only upon the first of these causes, presumably because the hypothetical sticky substance proved to be unnecessary. After considerable discussion of the 'horror of the void' with respect to phenomena involving gross bodies, the discussants turn to the properties of the particles composing these bodies. Salviati argues that it is the repugnance to the void which must "exist and be the cause of coherence between smaller parts, right on down to the minimum ultimate particles (*minime ultime*) of the same material. (Galileo, 1964–66, viii, p. 66; Galileo, 1974, p. 26.) It is not the resistance to the *formation* of a vacuum, but the *existence* of vacua between the particles of a solid which accounts for cohesion, as the ensuing reference to melting makes clear, though Galileo expresses himself in a very tentative manner. (Galileo, 1964–66, viii, pp. 66–67; Galileo, 1974, pp. 27–28.) He thus makes a distinction crucial for the understanding of his matter theory: a fluid, like a solid, resists the formation of a vacuum, (Galileo, 1964–66, viii, pp. 61–64; Galileo, 1974, pp. 22–24.) but unlike a solid, a liquid has no internal cohesion for it has no interparticulate vacua.[4] (Galileo, 1964–66, viii, pp. 103–108.)

The discourse seems to be interrupted at this point by lengthy digressions on geometry and infinity, and to this point Galileo's matter theory indeed seems similar to that of Hero. In fact, these 'digressions' are quite germane to his presentation, for they contain his highly original speculations as to the nature and number of vacua and the ultimate particles of solids. In the theory as presented thus far, the tiny voids must be sufficiently numerous to act upon every particle of the solid – but now many? Salviati implicitly raises this point, remarking that although these voids are tiny, the effect of an "innumerable multitude of them multiplies the resistances (to penetration or fracture) innumerably." (Galileo, 1964–66, viii, p. 67; Galileo, 1974, pp. 27–28.) Sagredo concurs, using the analogy of ants unloading a ship filled with grain, that the number of interparticulate vacua must be large but, of course, finite. Galileo here breaks with classical atomic theory, for he replies through Salviati that a finite solid can contain infinitely many voids and, by inference, infinitely many particles! (Galileo, 1964–66, viii, pp. 67–68; Galileo, 1974, p. 28.) This is the crux of his solution to the problems of explaining contraction and rarefaction in corpuscular terms: he proposes the existence of an infinite number of 'particles' within a finite body. The geometrical explication which then ensues of the paradox known as the *rota Aristotelis* (see Figure 1) is, in this context, intended as more than a refutation of the Aristotelian dictum that nothing continuous can be composed of indivisibles and that if something continuous is composed of points, these points must either be continuous or in contact with one another. (*Physica*, 231[a], 24–30) It is also a closely reasoned explication of Galileo's views on the nature of matter. When Galileo through Salviati speaks of points, he is also speaking of particles; of gaps or spaces, voids; and of lines or surfaces, bodies.

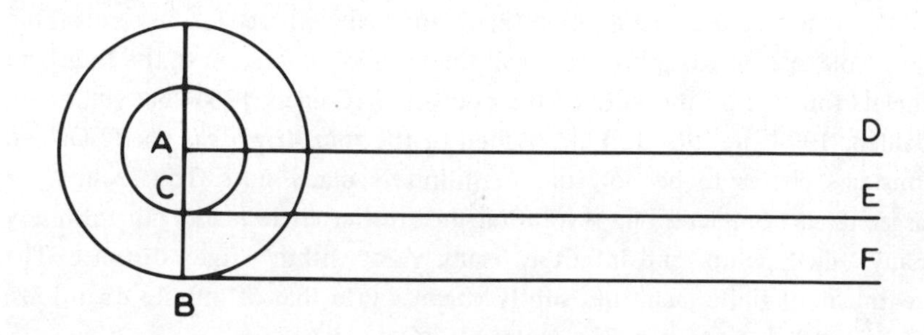

Fig. 1

Salviati proposes to solve this famous paradox first by considering a simpler problem involving two concentric hexagons. (See Figure 2) If hexagon *ABCDEF* "rolls" so that side *BC* fits the equal line *BQ*, side *IK* of the smaller

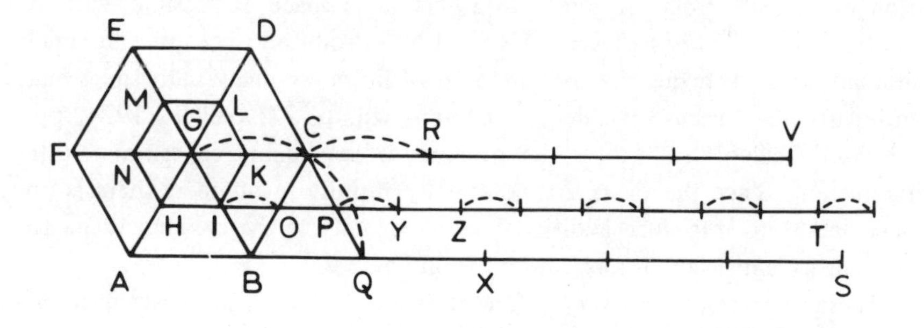

Fig. 2

hexagon will rest on line *OP*, having skipped over the segment *IO* without touching it. After one complete revolution of the larger hexagon, it will have marked along *AS* six contiguous lines, the total length of which is equal to the perimeter of *ABCDEF*. The small hexagon *HIKLMN* will also have touched along *HT* six lines equal to its circumference, but broken by five gaps. (Galileo, 1964–66, viii, pp. 68–69; Galileo, 1974, pp. 29–30.) Salviati claims that the same argument would apply to other polygons similarly arranged; *e.g.*, if a polygon of a thousand sides were to make one revolution, the smaller polygon within would mark out a thousand segments each equal to the length of one side and a thousand spaces between the segments.[5] It is evident that Galileo is not conducting a purely mathematical exercise, for he terms these gaps 'void spaces' since "we may call these 'void' in relation to the thousand linelets touched by the sides of the polygon." (Galileo, 1964–66, viii, p. 70; Galileo, 1974, pp. 30–31.) What then of the *rota Aristotelis* itself? Galileo considers circles to be polygons of infinitely many sides. Thus, when the larger circle completes one revolution the smaller circle marks out infinitely many 'filled points' and infinitely many voids within a finite distance. The treatment of finite quantities subtly changes into that of infinite quantities. (Galileo, 1964–66, viii, p. 71; Galileo, 1974, p. 33.)

Salviati here introduces a distinction crucial for the extension of this geometrical analysis to matter: the line traced out by the inner circle is not to be thought of as composed of parts which are countable, but of unquantifiable parts; that is, of infinitely many indivisibles. Thus, bodies are composed of infinitely many unquantifiable 'atoms' and, by inference, an infinity of unquantifiable voids. Salviati claims that from this point of view the expansion of 'a little globe of gold into a very large space' is possible without contradiction. No *quantifiable* voids need be introduced since even the small amount of gold in question is composed of infinitely many indivisibles and interparticulate vacua. (Galileo, 1964–66, viii, p. 72; Galileo, 1974, pp. 33–34.) Galileo thus avoids the problem of explaining cohesion despite rarefaction. Since the vacua are potentially infinite in number, there is no need either to introduce additional vacua or enlarge the existing vacua to account for expansion or rarefaction of solid bodies.

These conceptions are very different from those of classical atomism, as the ensuing 'digression' on the topic of infinity makes clear. By 'quantifiable' Galileo means 'countable', so that 'quantifiable voids' or 'quantifiable parts' must be understood to have dimensions. They are divisible. They cannot be mathematical points. Galileo's 'indivisibles' or 'unquantifiable parts' are uncountable. They are individually identical to mathematical points *but* can exist only in aggregates composed of an infinity of these points. These unquantifiable parts or indivisibles are not to be confused either with classical atoms which were particles incapable by their very nature of being divided or with the *minima naturalia* which were indivisible physically but divisible mathematically. Galileo's ultimate 'particles' are divisible neither physically nor mathematically because they are infinitely small; *i.e.*, mathematical points. He does, however, identify these unquantifiable indivisibles as 'atoms'. (Galileo, 1964–66, viii, p. 85; Galileo, 1974, p. 47.) The apparently unrelated ruminations upon the 'equivalence' of a point to a line, orders of infinity, the division of lines into quantifiable parts, and the 'unity of infinity' are a vital part of his effort to reconcile the geometrically possible; *i.e.*, infinitely small parts, with the dictates of common sense; *i.e.*, that an infinitely small non-extended 'atom' has no meaning. (Galileo, 1964–66, viii, pp. 73–85; Galileo, 1974, pp. 34–47.)

Galileo's examination of the different states of matter represents this conjunction between the mathematical and the physical. A solid body may be crushed into a fine powder, just as a line may be mechanically divided into a multitude of parts, yet neither process of division reaches its limit. The body cannot be crushed into indivisibles nor the line mechanically divided into points. The finest powder is still not a true liquid. The particles of a powder are still quantifiable and divisible. The powder may be heaped up or a depression formed in it. A liquid is composed of unquantifiable indivisibles with no voids interspersed, like the line resolved into an infinity of contiguous points. In Peripatetic terms the potential infinitude of parts in a solid cannot be actualized by mechanical division. Through heating, however, the vacua between the parts of a solid are filled with fire particles. This enables the solid particles which are potentially divisible into unquantifiables to become actually so divided. The result is the loss of internal cohesion and the transformation of the solid into a liquid. To return the fluid to the solid state, the fire particles are removed thereby restoring the interparticulate voids. Water, air, and similar substances have no interparticulate vacua, exist as aggregates of unquantifiable indivisibles, and are therefore penetrable.[6] (Galileo, 1964–66, viii, pp. 85–86; Galileo, 1974, pp. 47–48.)

Galileo does not deal with several problems which would seem obvious at this point in the *Discourses*. Why is gold, for example, naturally a solid or water a liquid? Why do oil and water, both liquids and therefore both continuous, have very different properties? In the former case, one may conjecture that gold 'naturally' contains considerably fewer fire particles than water or even none at all. It would seem, however, that an infinite number of fire particles would be required to melt a solid since it would contain potentially an infinity of void spaces to be filled. The term 'considerably fewer' thus seems rather inappropriate. In the latter case, since both oil and water are continuous; that is, made of an infinity of indivisibles with no interparticulate vacua, they possibly could be differentiated on the basis of the nature of the particles themselves. Galileo refers to 'odorous atoms' and the 'minute atoms' of a die, and implies the existence of atoms of fire, water, light, mercury, and other substances.[7] (Galileo, 1964–66, viii, pp. 85, 105, 134; Galileo, 1974, pp. 47, 64, 92.) Superficially it would seem that all infinitely small parts would be identical, but Galileo does point out that there are differing orders of infinite aggregates; *e.g.*, the set of all integers *vis-à-vis* the set of all primes. In any case he does not enter into any detailed explication of his apparent belief that the nature of a substance is to some extent determined by the nature of the particles which compose it.

Galileo's interpretation of expansion and contraction is the culmination of his disquisition on 'mathematical atomism'. Having treated expansion by means of the *rota Aristotelis* earlier, he uses this same paradox to explain condensation. This time the inner circle is rotated so that it marks out a line equal to its perimeter. (See figure 3.)

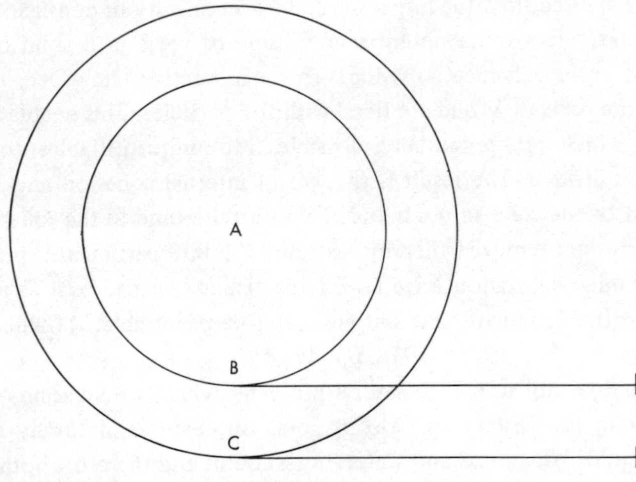

Fig. 3

How is the motion of the outer circle to be accounted for? Once more polygons are substituted for circles. (See Figure 4.)

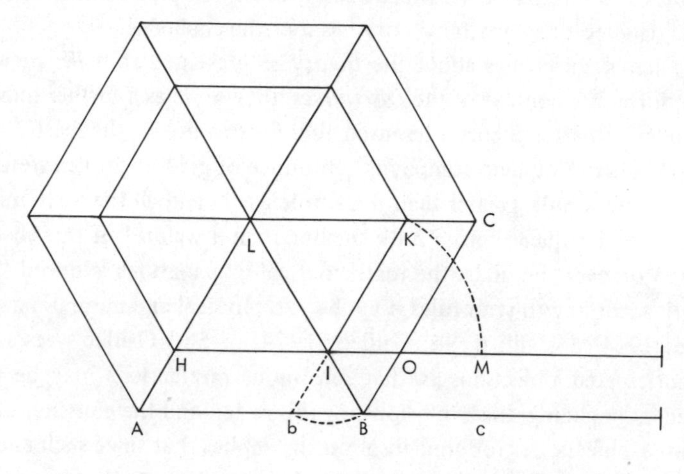

Fig. 4

If the smaller polygon *HIK* rotates, *K* traces out arc *KM*, and side *KI* will coincide with *IM*. *B* on the polygon *ABC* passes through arc *Bb* under line *ABc*, meanwhile, so that *BC* coincides with *bc* minus the part subtended by arc *Bb*, which is superimposed on the line *AB*. If the smaller polygon completes one revolution in this fashion, the larger polygon will trace out a line shorter than its perimeter by one less length *bB* than the number of its sides. Galileo then again treats the circle as a polygon with infinite and indivisible sides:[8]

> ... the infinitely many indivisible sides of the larger circle, with their infinitely many indivisible retrogressions, made in the infinitely many instantaneous rests of the infinitely many sides of the smaller circle, together with their infinitely many advances, equal to the infinitely many sides of the smaller circle, compose and mark a line, equal to that described by the smaller circle, ... making a compacting and condensation without any interpenetration of quantified parts. (Galileo, 1964–66, viii, pp. 93–95; Galileo, 1974, pp. 54–57.)

Galileo through his mathematical atomism avoids the interpenetration of parts which would occur were bodies formed of contiguous finite atoms compressed. The compression or condensation of a line is accomplished in the above analysis without any overlapping or interpenetration of parts, a result which is not obtainable with polygons since they have quantifiable parts; *i.e.*, sides. A solid body, since it is formed of potentially an infinite number of indivisibles and interparticulate vacua can thus be compressed in an analogous fashion.[9] Galileo's new matter theory has met the challenge.

'Simplicio's' misgivings about the theory as presented thus far are worthy of note, for in the context of the *Discourses*, they serve as a further indication that Galileo's matter theory presented therein was not in the least a mathematical whimsy. Simplicio complains "an ounce of gold might be rarefied and expanded into a bulk greater than the whole earth, and all the earth might be condensed and reduced into a bulk smaller than a walnut" if this theory be correct. Moreover, he adds, the mathematical explanations adduced "would not work according to your rules if applied to physical and natural materials." (Galileo, 1964–66, viii, p. 96; Galileo, 1974, p. 58.) Galileo's response to these anticipated objections is that enormous rarefactions may be readily observed as in gilding, the explosion of gunpowder, and the emission of odors by scented objects. As for condensation, he replies that since such enormous expansions can be observed, then reason dictates that equally enormous condensations must also be possible. (Galileo, 1964–66, viii, pp. 97–99, 104–105; Galileo, 1974, pp. 58–60, 64.) These responses to 'Simplicio's' charges reinforce the interpretation that though the discussions of matter theory are presented mathematically and do answer certain philosophical questions, at the same time Galileo is attempting to construct a plausible physical scheme.

The mathematical atomism set forth in the *Discourses on Two New Sciences* was indeed a most ingenious matter theory. Philosophically the theory satisfied Galileo's beliefs that the language of nature was mathematical and that it was the duty of the philosopher to seek out the mathematical, objective reality which underlay surface appearances. Physically, the twin problems of explaining rarefaction without sacrificing cohesion and of explaining condensation without surrendering the notion of impenetrability are, as Galileo himself puts it, "cleverly avoided by assuming the said composition of indivisibles." (Galileo, 1964–66, viii, p. 93; Galileo, 1974, pp. 54–55.) Although this theory seemed so satisfactory, Galileo presented it only as a tentative one, perhaps as a bold conjecture to stimulate further thought. The theory is but a sketch, with much detail missing and many problems unresolved.[10] Indeed, Galileo seemed to waver on such a fundamental question as whether the interparticulate voids – and by implication the atoms themselves – had dimensions after all rather than 'existing' as mathematical points. (Galileo, 1964–66, viii, pp. 105–106; Galileo, 1974, pp. 64–65.) Despite such uncertainties and ambiguities, he reduces the beauty and variety of nature to collections of atoms and voids. The demands of the new philosophy are fulfilled. As with so many of his other ideas, his matter theory drew upon the old and the new, the result being neither fully ancient nor fully modern, but certainly fully original and provocative.

University of Melbourne

NOTES

[1] For a brief discussion of the revival of atomism, see Hall (1949) and Hall (1952, pp. 413–433).

[2] The attribution of a Democritean atomism to Galileo on the basis of his distinction between subjective and objective qualities is extremely suspect. Aristotle himself made a similar distinction between 'common sensibles' and 'special sensibles' in his *De anima* (424^a–424^b). As for the references in *Floating Bodies*, it is true that Galileo criticized his Aristotelian opponents and had kind words for *some* of Democritus' ideas, but it is equally true that he rejected some of Democritus' assertions such as the suggestion that the floatation of a plate slightly denser than water was due to fire atoms.

[3] Galileo writes (Galileo, 1964–66, Vol. xvi, pp. 162–63) in 1634 of a discussion on these topics at Salviati's villa about eighteen years earlier, or 1616. Since Salviati died in 1614, the earlier date of 1614 would appear likely.

[4] Paul-Henri Michel's treatment of Galileo's views on the nature of matter (1964) is generally sound. However, he apparently understands those views to involve the existence of voids in *all* substances.

[5] As Drake comments in nn. 12, 13, and 14, Galileo glosses over the exact number of skipped spaces.

[6] Galileo elsewhere in the *Discourses* (Galileo, 1964–66, viii, pp. 114–116; Galileo, 1974, pp. 73–75) returns to a consideration of the cause of flotation, a subject he had treated some twenty-five years earlier in *Floating Bodies*. His newly-developed matter theory now enables him to explain why water is penetrable and why therefore its presumed resistance to penetration cannot be the cause of flotation. He still cannot, however, explain how drops of water can sustain themselves "as we see especially on cabbage leaves"!

[7] Galileo's reference to the die is rather curious. One could draw the inference from his treatment that the 'atoms' of a die are tiny cubes rather than mathematical points.

[8] Galileo here moves illegitimately from very small sides for polygons to point-sides for circles without in any way altering his argument.

[9] Since the composition of this paper, two related articles have come to hand. Smith (1976) provides an interesting study of parts of the First Day. Zubov (1964) is in agreement with the proffered interpretation of Galileo's geometrical arguments.

[10] For instance, he does not treat the expansion of fluids. On the basis of his explanation of the expansion of solids, one might argue that the infinitely small 'particles' forming liquids or air are in theory potentially further divisible and thus expansible. Alternatively, the addition or removal of fire particles could account for both expansion and contraction. It would seem that Galileo was attempting more to show the potential of his atomic hypothesis rather than to provide a complete system.

BIBLIOGRAPHY

Drake, Stillman and O'Malley, C. D., 1966, *The Controversy on the Comets of 1618*, University of Pennsylvania Press, Philadelphia, Pa.

Galileo, Galilei, 1964–66, *Le Opere di Galileo Galilei* (ed. by G. Barbèra), 20 vols. in 21 rept. of Edizione Nazionale, ed. by A. Favaro; Firenze.

Galileo, Galilei, 1974, *Two New Sciences* (transl. by Stillman Drake), University of Wisconsin Press, Madison, Wisconsin.

Hall, Marie Boas, 1949, 'Hero's *Pneumatica*: A Study of its Transmission and Influence', *Isis* 40, 38–48.

Hall, Marie Boas, 1952, 'The Establishment of the Mechanical Philosophy', *Osiris* 10, 412–541.

Lasswitz, Kurd, 1890, *Geschichte der Atomistik vom Mittelalter bis Newton*, 2 Vols., Hamburg.

Löwenheim, Louis, 1894, 'Der Einfluss Demokrit's auf Galilei', *Archiv für Geschichte der Philosophie* 7, 230–268.

Melsen, Andrew G. van, 1952, *From Atomos to Atom* (transl. by Henry J. Koren), Duquesne University Press, Pittsburgh.

Michel, Paul-Henri, 1964, 'Les notions de continu et de discontinu dans les systèmes physiques de Bruno et de Galilée', *Mélanges Alexandre Koyré*, vol. 2, Hermann, Paris, pp. 346–359.

Shea, William R., 1970, 'Galileo's Atomic Hypothesis', *Ambix* **17**, 13–27.

Smith, A. Mark, 1976, 'Galileo's Theory of Indivisibles: Revolution or Compromise?', *Journal of the History of Ideas* **37**, 571–588.

Zubov, Vassily P., 1964, 'Atomistika Galileia', *Voprosy Istorii Estestvoznaniia i Tekhniki* **16**, 38–51.

ERNAN McMULLIN

THE CONCEPTION OF SCIENCE IN GALILEO'S WORK*

It has been remarked more than once that each generation of theorists of science makes of Galileo, the "father of modern science" by customary reckoning, a scientist after its own heart.[1] Most recently Paul Feyerabend proposed that:

> What Galileo did was to let refuted theories support each other, that he built in this way a new world-view which was only loosely (if at all!) connected with the preceding cosmology (everyday experience included), that he established false connections with the perceptual elements of this cosmology which are only now being replaced by genuine theories (physiological optics, theory of continua), and that whenever possible he replaced old facts by a new type of experience which he simply invented for the purpose of supporting Copernicus.[2]

The reason why Galileo had to resort to "propagandistic machinations" and "trickery" of all sorts was that argument alone would not have sufficed to make his case.[3] What happened was that in his Copernican campaign, experience ceased:

> to be the unchangeable fundament which it is both in common sense and in the Aristotelian philosophy. The attempt to support Copernicus makes experience "fluid" . . . An empiricist who starts from experience, and builds on it without ever looking back, now loses the very ground on which he stands. Neither the earth, "the solid, well-established earth", nor the facts on which he usually relies can be trusted any longer.[4]

Feyerabend's attack on empiricism rests on a detailed analysis of the argument of the *Dialogo*, which he takes to be a paradigm of scientific persuasion generally.

This construal of Galileo finds an echo in a well-known passage in the *Critique of Pure Reason*, written almost two centuries earlier:

> When Galileo caused balls, the weights of which he had himself previously determined, to roll down an inclined plane. . .a light broke upon all students of nature. They learned that reason has insight only into that which it produces after a plan of its own, and that it must not allow itself to be kept as it were, in nature's leading-strings, but must itself show the way with principles of judgment based upon fixed laws, constraining nature to

R. E. Butts and J. C. Pitt (eds.), New Perspectives on Galileo, 209–257.

give answer to questions of reason's own determining. Accidental observations made in obedience to no previously thought-out plan, can never be made to yield a necessary law, which alone reason is concerned to discover. Reason, holding in one hand its principles, according to which alone concordant appearances can be admitted as equivalent to laws, and in the other hand the experiment which it has devised in conformity with these principles, must approach nature in order to be taught by it. It must not, however, do so in the character of a pupil who listens to everything that the teacher chooses to say, but of an appointed judge who compels the witnesses to answer questions which he has himself formulated.[5]

Kant and Feyerabend agree on the "fluidity" of the experience on which science rests, on the active role of the scientist in superimposing on its structures of intelligibility of his own fashioning. They would agree too that the scientist is not entirely free in this work of construction; a balance of sorts has to be reached:

While reason must seek in nature, not fictitiously ascribe to it, whatever as not being knowable through reason's own resources has to be learnt, if learnt at all, only from nature, it must adopt as its guide, in so seeking, that which it has itself put into nature.[6]

Or as Feyerabend puts it, allegiance to a new theory has:

to be brought about by irrational means such as propaganda, emotion, *ad hoc* hypotheses, and appeal to prejudices of all kinds. We need these "irrational means" in order to uphold what is nothing but a blind faith until we have found the auxiliary sciences, the facts, the arguments that turn the faith into sound "knowledge".[7]

The two disagree fundamentally, of course, about the status of science as "necessary laws," about its grounding in the permanent and universal structures of the human reason. But what interests us here is their unanimity in seeing Galileo as the prototype of the scientist whose task it is to lead people to "see" Nature as he already, even in advance of the "facts," has come to see it.

It is hardly necessary to say that this account of Galileo is diametrically at odds with the one still found in most physics textbooks, and in inductivist accounts of science from the eighteenth century onwards. Whewell expresses this tradition very well when he sums up Galileo's approach by saying that he:

had perhaps a preponderating inclination towards facts, and did not feel, so much as some other persons of his time, the need of reducing them to ideas.[8]

Mach, characteristically, gives us the seeker after laws patiently rolling balls down inclined planes to test out plausible law-like hypotheses formulated in advance:

> Galileo did not supply us with a *theory* of the falling bodies, but investigated and established, wholly without preformed opinions, the *actual facts* of falling. Gradually *adapting* ...his thoughts to the facts, and everywhere logically abiding by the ideas he had reached, he hit on a conception which to himself, perhaps less than to his successors, appeared in the light of a new law.[9]

One may be tempted to remark indulgently at this point that Galileo has evidently become all things to all men, and go on perhaps to make the Feyerabendian point that this illustrates anew the ability possessed by protagonists of a theory (whether of Nature or of a historical figure) to superimpose their own lineaments when "discovering" intelligibility in their material. It does seem, however, that a divergence of interpretations as deep as this one may derive in part from something in the material itself. Why is it that the battle focusses around *Galileo*, and not, for example, around Boyle or Lavoisier or Darwin? Is it only because he is the "father" of modern science, and thus to be appropriated at all costs by philosophers who would propose an account of how science is carried on?

In this essay on Galileo's theory of science, we will try to answer this question. We will see that Galileo's science is a diverse enterprise, pursued in many different contexts, following methods which altered over the years. More fundamentally, we will see that he was the inheritor of a strict notion of science as demonstration, but of quite ambiguous views as to how this demonstrative character is to be achieved. It is risky to take his work (above all, his efforts on behalf of Copernicanism) as a paradigm of science because his view of science as necessary demonstration leans far more to the ancient world than to our own.

1. AN AMBIGUOUS HERITAGE

Galileo's attempt to construct, not only a new science, but a new account of the nature of scientific inquiry itself has to be seen against the background of the Aristotelian view he sought to replace. Instead of surveying this latter as a whole,[10] we shall concentrate just on two general features of it that are indispensable to an understanding of the problems Galileo faced.

The sharp distinction Aristotle drew between physics and mathematics left him in a dilemma about such areas of inquiry, already flourishing in his day, as optics, harmonics, mechanics, and above all, astronomy. For each of these made extensive use of mathematical methods and mathematical language in their efforts to understand some part of the physical world. Aristotle's own discussion of the rainbow in the *Meteorology*[11] relies on geometrical analysis; in the *Physics*, he is forced to make use of mathematical principles in his treatment of such key notions as infinity, continuity, and velocity. Even the very instances of eclipse and lunar illumination he gives in the *Posterior Analytics* to illustrate the nature of scientific demonstration rely implicitly on geometrical analysis.

In the circumstances, one might have expected him to say that these sciences constitute a "mathematical physics," or even to admit that the language of at least some parts of physics is *necessarily* mathematical in form. But the barriers against such a move were much too high. Plato had been willing to make it, though sceptical about the possibility of a proper "science" of physics in the outcome. But Aristotle, the biologist, had carefully constructed an account of nature and causal explanation where the qualitative took priority over the quantitative, and where the laying out in syllogistic order of essences and properties would be hindered rather than helped by focussing on the geometrical aspects of nature only. And so he hesitated between making these sciences "the more physical part of mathematics,"[12] and allowing them a special sort of intermediate status between physics and mathematics, depending for their scientific character on principles borrowed from mathematics.[13] Both suggestions were taken up in the later tradition. Astronomers and theorists of light were usually described as "mathematicians," though they themselves often warned against the implication that they were not concerned with physics.[14] An elaborate theory of "middle sciences" was worked out in medieval philosophy which made mathematized mechanics an intermediate science subalternate to mathematics and distinct from physics. The aim in part was to retain for physics the basically non-mathematical structure Aristotle seemed to have imposed upon it. The story of these *scientiae mediae* in the later Aristotelian tradition is an extremely complex one and it would take us too far afield to dwell on it in the detail it derserves.[15]

One thing is clear, however. There were grounds in Aristotle's own work for wondering whether a physics without mathematics is possible. But as time passed, the separation between the ordinary-language analysis characteristic of Aristotelian natural philosophy and the complex mathematical and observational techniques of the astronomers and optical theorists made collaboration (or even contact) between the two professional groups very difficult. Nonetheless, the question faced each generation of natural philosophers anew: if mathematics is integral to optics, mechanics, astronomy, how can a demonstrative science of physics be constructed without its aid?[16]

A second ambiguity of the Aristotelian theory of science is of even more concern for our story. In the *Posterior Analytics*, Aristotle specifies the kind of "showing" (*apodeixis*) that would, in his view, qualify as "science" (knowledge in the fullest sense, *epistēmē*). Now 'showing' in Greek (as in English) has three rather different senses: to show may mean to *prove*, or to *explain*, or to *teach*. Aristotle weaves all three senses together in his account of *apodeixis* (usually translated in English as "demonstration," a defective rendering because the sense of *apodeixis* as explanation is not conveyed). For knowledge to qualify as *fully* "scientific," it must, it would seem, fulfill all three goals.

The paradigm of science for Aristotle was geometry, of course, as it had been for Plato. Geometry was known to provide knowledge in the strongest imaginable sense of that term: necessary and eternal truths. It was natural, then, to take geometrical procedures as a model for scientific procedures generally. This suggested that science should rest on axioms or premisses, themselves self-evident, from which theorems or conclusions could be derived. The self-evidence of the premisses could then be transferred, as it were, to the conclusions. Two movements of thought would be required: one of "induction" (*epagōgē*) by means of which one would hit upon the appropriate premisses and grasp their necessity, the other that of deduction, by means of which one would move by secure logical rule from premisses to conclusions.

In geometry, it was easy enough to see how this would work, and how the "showing" would be proving, explaining, and teaching, all at once. The axiom-theorem relationship proves the theorem precisely by giving the reason for (i.e. by explaining) it. But in natural science, it was not as obvious how this would be done. These were two immediate difficulties. First, how would this "induction" work, and how could it possibly guarantee the *necessity* of the connections alleged? It was all very well to do this in the realm of lines and points, but organisms and planets are something else entirely. Aristotle asserts that induction begins from sense-perception, discovers unity in the memory of successive perceptions and is thus enabled to abstract a universal. The faculty whereby man can form such universals and grasp the connections between them in an infallible one, *nous*. The exercise of *nous* in the formulation of first principles by means of *epagōgē* is prior to (and a necessary condition of) the *apodeixis* that constitutes science.[17] But can one *really* hit upon a premiss like "all planets are near" (his example) quite so simply, and with absolute certainty of its truth?

The second difficulty was with the premisses. Were they to be construed as first principles, very general axioms, as in the geometrical paradigm? Or were they to be construed as statements of proximate causes, as in practice Aristotle more often appears to assume? He writes as though the two formulations are equivalent, or at least easily reconciled by postulating a hierarchy of causes leading back to first causes, grasped in the first principles of the science. But it is by no means clear that such a hierarchy exists in natural science (as reference to his own examples of non-twinkling planets and lunar eclipses shows), such that the entire array of lowest-level propositions are deductively derivable from a small number of very general axioms. The problem here once again was with the use of an axiomatic science like geometry as a model for physics or biology.

What made both these difficulties even more intractable was the modification Aristotle introduced into this analogy, making science not only axiomatic but also syllogistic. His reasons for doing this were rooted deeply both in his metaphysics and his biology; they are quite complex and need not detain us here. But now the tension with geometry (whose connectives are not in the least syllogistic) is obvious. *Apodeixis* is to be construed as the locating of properties in essence, and the situating of the properties relatively to one another. It is also the interrelating of essences by generic and differentiating notes. But how are the "first principles" to be arrived at here? How does induction operate? How can it provide knowledge of truly essential characteristics? Above all, how are causal relationships discovered, and where do they fit into the scheme?

Aristotle tells us that man has a gift of "quickness of wit" (*achinoia*) which enables him to hit upon the middle (causal) term linking extremes. If the problem is: why does the bright side of the moon always face the sun?, the answer is: because the brightness of the moon is caused by the sun. Or in the cumbrous syllogistic mode:

Perception of the extreme terms enables him to recognize the cause or middle term. *A* stands for "bright side facing the sun", *B* for "deriving brightness from the sun", and *C* for "moon". Then *B* applies to *C* and *A* applies to *B*. Thus *A* applies to *C* through *B*.[18]

But what sort of quick wit is this that allows one to discover and assert with absolute assurance that whatever has its bright side facing the sun derives its brightness from the sun? Must one not have a fairly elaborate theory of light, and must one not also make use of geometrical analysis? Can this premiss *really* be based on a sort of induction-by-abstraction?

In one of the most commented-upon passages in all of Aristotle's work, this difficulty shows itself even more clearly. Aristotle distinguishes between two types of scientific knowledge (*epistēmē*), one of the "what" (*oti, quia*), the other of the "why" (*dioti*, *propter quid*). One deductively establishes facts on the basis of other known facts (e.g. observational astronomy), the other "gives the why of" the facts (e.g. mathematical astronomy). Since an *apodeixis* of the *oti* must begin from the "familiar," and the familiar may well be the effect rather than the cause, it will in such cases work back from effect to cause, and the "what" established in the conclusion will be the cause.[19] Thus we know that planets do not twinkle, and (by induction or perception, he says) that non-twinkling things are near; we can thus conclude that planets are near. The *apodeixis* here is a means of *proving* (establishing the truth of) the conclusion, not of explaining it. Its function would be one of *discovery* also, since we go from what we know well to what we do not know; by our quickness of wit, we have perceived the requisite middle term (non-twinkling).

A very different sort of *apodeixis* would be one which would conclude that planets do not twinkle because they are near. There is no question here of proving the *truth* of the conclusion from the premisses, since it is in fact better warranted than they, to begin with. Rather, the premisses *explain* the conclusion. They do so, not just because of the soundness of the syllogism linking planet, nearness, and non-twinkling, but because nearness is perceived to be the *cause* of the non-twinkling. How is this known? The truth of the premiss: nearby things do not twinkle, is not sufficient to establish a causal relation. Nor is it sufficient (as Aristotle at times appears to assume) that the terms of the major premiss are convertible, that nearness and non-twinkling are mutually predicable. Invariable co-presence of *A* and *B* does not tell us whether one *causes* the other. This problem would not arise in geometry. But in physics, there will have to be some way to decide on causal order and relation among properties; Aristotle does not give much help on this.

Apodeixis has thus two separate functions in natural science, unlike its role in mathematics where no such duality is found. The reader is left in no doubt as to which of the two is the more basic: "the most essential part of knowledge is the study of reasons (*dioti*)."[20] If the cause happens to be better known than the effect, the same form of *apodeixis* will accomplish both ends together; it will *explain* and *establish* the conclusion simultaneously. But if it is not (e.g. if the nearness of the planets is less familiar than their non-twinkling), then the task of the scientist splits into two, one to discover and establish the "what" (the nearness of the planets), the other to show that this is, in fact, the cause of the property to be explained (their non-twinkling).

But are they really separate tasks? Could one establish the "what" without also knowing the "why"? Could one establish the nearness of the planets without knowing that distance is the cause of twinkling in planets? Could one explain the non-twinkling character of planets unless one already knows it to be the case that they are near? And is there not a disturbing circularity about this procedure, if both directions are taken? If planets are not *known* to be near, then one cannot explain (in Aristotle's sense of *apodeixis*) why they do not twinkle. And if nearness is not *known* to be the cause of non-twinkling in planets, it is not clear whether one could establish (in the other sense of *apodeixis*) the fact that planets are near.

The problem lies here with Aristotle's first making cause a relation between properties, and then assuming that a syllogistic of such relations can provide a set of necessary and certainly-known truths. His optimistic suggestion is that one can ascend from the familiar sight of planets that do not twinkle, via the observation that non-twinkling things are always near, to the certain knowledge that planets are near. If the major is convertible, i.e. if nearby things do not twinkle, *and* if a causal relationship between distance and twinkling can somehow be "seen," then a *conclusive* explanation of the twinkling of planets can also be arrived at. No wonder later Aristotelians created such an elaborate structure around this sketchy account of ascent and descent in an attempt to circumvent the obvious difficulties.

It was in keeping with the thrust of Aristotle's distinction between the functions of proof (*oti*) and explanation (*dioti*) to think of the two as complementary, since neither seemed to stand firm in its own right. A variety of distinctions (between analysis and synthesis, between resolution and composition) were later added to the original Aristotelian one to create a literature of fearsome complexity in which these distinctions varied in meaning from one writer to the next. ('Analysis', for example, sometimes implied the ascent from effect to cause, sometimes the descent from means to end). It was in the Paduan school of the sixteenth century that this literature reached its greatest elaboration, in the work of Zabarella notably, in his theory of a "*regressus*" that would combine in a single demonstrative whole the movements of ascent and descent.

We shall not attempt to recreate this discussion here.[21] The reason for this long excursus is simply to indicate what some of the main weaknesses of the Aristotelian notion of *apodeixis* were in the context of natural science, weaknesses that generations of commentators and natural philosophers puzzled over. We are concerned not to trace Galileo's immediate antecedents in the contemporary manuals he read, but rather to see his theory of science against the broad background of the *Posterior Analytics* he knew so well.[22] It seems reasonable that he should be situated by reference to the theory of science he strove to replace and yet to which he remained in many ways so closely bound. Though it was his debt to Plato and Archimedes that he himself preferred to stress, the climate in which its notions of science and demonstration were formed was basically Aristotelian. And, as we shall see, the difficulties his account of science left unresolved were ones that had dogged the scientific enterprise from its beginnings in Greek thought.

2. GALILEO'S PROGRAM IN THE *DISCORSI*

When Galileo in the Discorsi proposed a "brand-new science concerning a very old subject", what made it "brand-new" in his eyes was not primarily that it followed a new method or propounded a new conception of science, but rather that it revealed features of motion that "had not hitherto been remarked, let alone demonstrated."[23] It was fashionable in his day to assert that the methods of natural science had to be reformed, that a new *Organon* had to be created which would allow a more penetrating and more technologically effective science to be constructed. But this was not the sort of emphasis that Galileo chose to give his work. His aim was to discover a demonstrative science of motion, not to write a *Discourse on Method.* His remarks on method are always elicited by the scientific problem immediately at hand;[24] one never has the impression that the problem is chosen with a view to illustrating a theory of method. In our terms, he reads as a scientist rather than a philosopher. His primary claim is to a new science, not to a new theory of science.

In this regard, the argument of the *Dialogo* is instructive. Galileo sets out to destroy Aristotle's physics, to show that it lacks demonstrative character and is even in some respects incoherent. But he is careful to exempt Aristotle's general conception of science from his criticism; it is precisely because he can claim to share a common theory of demonstration with his Aristotelian opponents that he can also hope to persuade them of his case. At one point, Salviati reassures Simplicio that their disagreement has not to do with "the manner of philosophizing," but rather with the consequences of "new events and observations, such that if Aristotle were now alive, I have no doubt he would change his opinion."[25] Simplicio responds by specifying what he takes Aristotle's "manner of philosophizing" to have been: conclusions are to be warranted first *a priori* on the basis of intuitively evident principles, and then *a posteriori* by reference to the evidence of the senses and to tradition.

Salviati retorts that this is the order of exposition not of discovery. Aristotle (he feels sure) arrived at his conclusions in the first place "by means of the senses, experiments and observations" just as he himself did. Only afterwards did he seek a way to derive them from already demonstrated premisses or from first principles. But whether or not he is correct in regarding Aristotle in this way, in order to show their basic agreement, it is enough, he says, "that Aristotle, as he said many times, preferred sensible experience to any argument."

What is important here is not whether Aristotle really *was* the empiricist the *Dialogo* makes him out to have been. What interests us is that Galileo nowhere suggests (and even goes to some pains to deny) that his disagreement with Aristotle extends to the conception of science itself. There is every reason to suppose that Galileo simply took for granted that science has, in fact, the logical structure Aristotle supposed it to have. Were this not to have been the case, he would assuredly have given us more clues than he did about a new and different conception of science.

Let us assume, then, that we can overcome this first temptation which is to claim that Galileo proposed a radically new notion of science. There is still a second temptation: to suppose that Galileo was in possession of a well-articulated coherent theory of science which he consistently employed throughout his entire scientific work. We shall see that this is not the case. The methodological hints that he throws out ought to warn us against any such assumption. The same Salviati who, for example, praises Aristotle for having "properly preferred sensible experience to natural reason" is equally enthusiastic about Aristarchus and Copernicus for having made reason "the mistress of their belief" in defiance of the senses.[26] The tensions within his conception of science have, as we shall see, quite easily traceable sources. It would have been very difficult, indeed, for him to overcome them.

His importance for posterity lay in the fact that he laid the foundations of a successful mechanics. It was for this reason, and not because of any formal theory of science he proposed, that the methodology underlying his work later took on a special authority. In this essay, we propose to discuss one specific aspect of Galileo's methodology, namely his conception of science. It will be important to investigate not merely his mechanics but also his work in astrophysics (where he explicitly seeks a causal understanding), and the debate on the Copernican question. Each of these three contexts presented him with a rather different challenge in regard to the sort of science apparently attainable there.

Let us be clear from the outset about what is meant by a "conception of science." Three questions will be asked. First, what is the status of the truth-claim made by the scientific assertion? Second, what mode of inference links premiss and conclusion? Third, how does experience serve to warrant the claims made? Between them, these specify a "conception of science." It will be found that two quite different conceptions are in open tension in Galileo's work. The tension itself was not a new one, but in Galileo's physics the issues were for the first time clearly joined.

3. DEMONSTRATIVE SCIENCE AND NECESSARY TRUTH

There was never any doubt in Galileo's mind as to the distinguishing characteristics of that knowledge which qualifies as "science":

If what we are discussing were a point of law or of the humanities, in which neither true nor false exists, one might trust in subtlety of mind and readiness of tongue and in the degree of experience of the writers, and expect him who excelled in those things to make his reasoning most plausible, and one might judge it to be the best. But in the natural sciences, whose conclusions are true and necessary and have nothing to do with human will, one must take care not to place oneself in the defence of error, for here a thousand Demosthenes and a thousand Aristotles would be unable to overcome a mediocre wit who just happened to hit upon the truth.[27]

Plausibility may be all very well in law or the humanities, but it will not do in the natural sciences where conclusions must be both true and necessary. Galileo praises Gilbert for the "stupendous concept" of magnetic force he formulated and "the many new and sound observations" he brought in its support. But he laments the lack of grounding in geometry which led Gilbert to be insufficiently critical of the reasonings to *verae causae* be employed. They "lack that force which must unquestionably be present in those adduced as necessary and eternal scientific conclusions."[28]

The terminology of "true, necessary, and eternal" is, of course, that of the *Posterior Analytics*. In his youthful commentary on that work, already referred to, Galileo stresses that true knowledge is of causes and is attained by "necessary demonstration," a phrase which will recur very frequently in his later writings.[29] In the physical sciences (where effects are usually better-known than causes), one has to work back to the causal premiss in demonstrative fashion. He assumes that the same effect cannot be brought about by a multiplicity of different causes; there is a unique relationship, traceable demonstratively, from effect to *vera causa*. In other words, all causal hypotheses save one can be excluded if the effect is properly understood. And once the causal premiss is reached, it can be seen to be "true and necessary" in its own right (otherwise the demonstrative reasoning from cause back to effect would be circular).[30]

The ambivalence of the Aristotelian account is familiar. What *is* the warrant of the causal premiss? It is supposed to be evident in its own right. Demonstrative order requires (he notes at the beginning of his *Discourse on Floating Bodies*) that the propositions from which the demonstration begins should themselves be "true and obvious."[31] He sometimes envisions these propositions as definitions, intuitively acceptable once understood:

That which must be observed in all the demonstrative sciences we must also carry out in this treatise; that is, to propound the definitions of the special terms of this study and its basic assumption; from which, as from fertile seeds, will germinate and spring as consequences the causes and true proofs of the properties of all mechanical instruments.[32]

This is how *On Mechanics* opens, and it goes on to develop in deductive form a whole series of "theorems," after the fashion of Archimedes. It is presumed that the starting points are unproblematic so that "certain demonstration" is possible.[33] They are, in fact, reasonable assumptions of a persuasive sort.

In the context of statics, such assumptions are not so difficult to come by. But can one count on intuitively evident principles in other domains? In mechanics, the principle that difference of weight has no effect on speed of fall "at first glance seems so remote from probability" that every effort must be made to find "experiments and reasons" to corroborate it.[34] And when specific causes are sought, one *may* be able in favorable cases to argue from effect to cause, but will the causal premiss then come to possess truth and necessity in its own right, so as to furnish the basis for strict demonstration? In the *Dialogue*, Galileo argues that the complex motions of the sunspots across the sun's disc can only be understood if one attributes a double motion to the earth. This "yields easily and clearly the true cause of such strange phenomena."[35]

Even assuming that this inference from effect to unique cause goes through, in what sense can the causal premiss, stating the double motion of the earth, be said to be "true and necessary"? It is certainly not necessary by any necessity of nature on the part of the earth. Nor is this premiss capable of standing as a first principle, seen to be true once properly understood. Galileo goes on to remind the reader that the motion of the earth is supposed to remain an open question, but leaves no doubt about his own view. He ends: "it is not possible that the reasons adopted by the right side should be anything but clearly conclusive."[36]

Suppose his argument were conclusive, would this make it demonstrative? Not if the term is used in the Aristotelian sense. It is not enough for the argument to yield a true conclusion; the premisses must also be necessary in their own right, either as first principles or as demonstrated from such principles. But where are premisses of this sort to be found? Only in mathematics, it might seem:

Taking man's understanding *intensively*, insofar as this term denotes understanding some proposition perfectly, I say that the human intellect does understand some of them perfectly, and thus in these it has as much certainty as Nature itself has. Of such are the mathematical sciences alone; that is, geometry and arithmetic, in which the Divine intellect indeed knows infinitely more propositions since it knows all. But with regard to those few which the human intellect does understand, I believe that its knowledge equals that of the Divine in objective certainty, for here it succeeds in understanding necessity, beyond which there can be no greater sureness.[37]

For strict demonstrations, necessary connections have to be grasped, and this, it would seem, is possible only in arithmetic and geometry. It is not enough that there should be a "necessary" (i.e. a valid deductive) connection between premisses and conclusion; if the conclusion is to be "eternal and necessary," so must the premisses themselves be, and such premisses require that necessary connections between the constituent concepts be comprehended. Aristotle believed the human mind to be capable of understanding the essences of natural things well enough to grasp the necessary connections of essence and property, in some cases at least. This would allow the scientific understanding of agent causes, where (but only where) the activity to be understood can be construed as a regular outcome of nature, as in the natural behavior of organisms, for example. The pattern here is of an agent's being observed to act in a certain way; one asks whether this activity derives from (is caused by) the agent's nature. But in many cases the cause is remote from human inspection or ambiguous in its operation. In a famous passage in *The Assayer*, Galileo illustrates the difficulty of tracing effect to cause, even in an apparently simple phenomenon like a cricket song. He concludes:

I might by many other examples make clear the bounty of nature in producing her effects by means which we would never think of if our senses and experience did not teach us of them, though even these are sometimes insufficient to remedy our incapacity. Therefore, I should not be denied pardon if I cannot determine precisely the manner in which comets are produced, especially as I never boasted that I could, knowing that it may occur in some way far beyond our power to imagine.[38]

Here it is the "bounty of nature" that prevents the easy tracing of effect to cause. In the *Discourse on Method*, Descartes claims that from "the general principles or first causes of all that is," he can derive a deductive science which will "discover the skies, the stars, an earth." But at a certain point, deduction fails because:

The power of nature is so ample and so vast, and these principles so simple and so general that I almost never notice any particular effect such that I do not see right away that it can be derived from these principles in many different ways.[39]

Only experiment, he concludes, can determine which of the ways is correct, which causal hypothesis is true.

The second difficulty is that the operation of the cause may be "beyond our power to imagine." Galileo often reflects on this:

It always seems to me extreme rashness on the part of some when they want to make human abilities the measure of what nature can do. On the contrary, there is not a single effect in nature, even the least that exists, such that the most ingenious theorist can arrive at a complete understanding of it.[40]

Much less, then, are his chances of understanding the causes that operate on the lunar surface; "if I were asked what my basic knowledge and natural reason told me regarding the production there of things similar to or different from ours, I should always reply: 'very different and entirely unimaginable by us'."[41] This is unduly pessimistic, given what Galileo has just had to say about the earth-like features of the lunar surface. But it is indicative of a realistic appraisal of the difficulties in the way of causal analysis of complex or remote natural processes.

Does this mean that Galileo abandoned the search for causes? Clavelin appears to think so. Indeed, he argues it to be a merit of Galileo that he moved science away from the search for causes, which had been such a "source of confusion" in Aristotelian thought. Making such a search "the aim of natural philosophy", he says, "means hampering the very formulation of scientific problems."[42] The argument of the early *De Motu* "was still conducted from a causal point of view." But in Galileo's later works, "a truly novel relationship between experience and scientific reason appeared," a new notion of explanation which "finds its true meaning in the anticipation of facts yet to be discovered," rather than in the description of causes.[43] In his analysis of the sunspot movements, Galileo "equated explanation with the expression of a phenomenon or a set of phenomena in a rational system and with the help of a model."[44] He appealed to the principles and concepts of this system "by the elaboration of a model designed to express the phenomena," thus "transforming simple matters of fact into objects of interpretation," uniting them in "a coherent whole governed by general principles and ideas."[45] The appeal to coherence and simplicity has thus (in Clavelin's view) replaced the search for causes in Galileo's cosmology.

In the *Discourses*, he went much further than that: here an explanation became the transformation, with the help of an appropriate construction, of a physical fact into a mathematical problem, followed by its analysis and solution in the light of previously established mathematical truths. Hence, though Galileo did not fundamentally change his explanatory ideal as he passed from his cosmological studies to the *Discourses*, it was only in the latter that this ideal came into its own and that physical research was set on the road it has been following ever since.[46]

This reading of Galileo has something of Descartes in it and prefigures much of Kant. It is not a good reading, as the remainder of this essay will attempt to show. In particular, Clavelin's assertion that Galileo turned away from causal explanation (one of "the two chief factors responsible for the impotence of the traditional method"[47]), and forged a single explanatory ideal for all of science, resolving physical problems in the light of mathematical principles, is open to serious challenge. Galileo did not turn (and could not turn) away from causes; his work exhibits more than one explanatory ideal; the subordination of mechanics to mathematics is not original with him, but is characteristic of the tradition of the *scientiae mediae*; it is not, in fact, the conception that distinguishes modern science, not even if one reduces science to mechanics (as proponents of this claim often tend to do). Causal explanation is as central to science today as it ever was; only in mechanics do perplexities arise about the causal status of such concepts as force and energy.

But Galileo is our concern here. And he speaks freely of "true, intrinsic and proper causes,"[48] and argues in the *Dialogue* that "knowledge of the effects is what leads to an investigation and discovery of the causes."[49] Even in the *Discourses*, after calculating a projectile trajectory, he adds: "The knowledge of one single effect acquired through its causes opens the mind to the understanding and certainty of other effects without need of recourse to experiments."[50] What concerns him most is that effects:

must not merely follow easily, they must follow necessarily, in such a way that it would be impossible for them to take place in any other manner. For such is the property and condition of things which are natural and true.[51]

His analysis of the tides is wholly causal in conception; it is only because the periods of the tides "depend upon invariable causes which are unified and eternal" that scientific conclusions of the requisite "fixed and constant" sort can be arrived at.[52] He indicates the method to be followed:

If I say that it is true that one effect can have only one basic cause, and if between the cause and the effect there is a fixed and constant connection, then whenever a fixed and constant alteration is seen in the effect, there must be a fixed and constant variation in the cause.[53]

Mill might not agree with the "one effect, one cause" restriction. But Galileo is trying to find a way to make the effect-to-cause inference a "necessary" one, as Aristotelians had tried to do with the demonstration *oti.* The necessity of such an inference lies in the "necessity" of the deduction and in the "necessity" of the causal relations presumed to underlie the premisses. Are these sufficient to constitute the conclusions as "necessary truths"? Galileo has no doubt that they are. But how is one to know that the premisses are true in the first place? Granted that if they are true, they are (in Galileo's account) necessarily true, can one grasp the concepts (or understand the causes) sufficiently well to assert that they are true? This is the ambiguity in which Galileo's conception of science constantly revolves.

This ambiguity, as Galileo well knows, does not arise in mathematics. And so there will be the constant temptation to reduce natural philosophy to mathematics, to make mathematical principles the sole source of physical truth. "The force of necessary demonstrations is full of marvel and delight; and such are mathematical /demonstrations/ alone."[54] The necessity of mathematical propositions can be immediately grasped without need of recourse to physical causes. The full demonstrative ideal of science, it is plain, can be achieved only to the extent that a physical science can simulate mathematics.

4. THE STATUS OF MECHANICS AS A SCIENCE

The language of the *Discourses* is dotted with terms like "demonstration" and "rigorous proof." The First Day opens with a promise to "prove demonstratively and not just persuade by probable arguments."[55] The propositions of both new sciences are to be "geometrically demonstrated"; this is what makes the sciences "new," because even those propositions that were previously known "were not proved by necessary demonstrations from their primary and unquestionable foundations," as they must be to constitute science in the strict sense. At the end of the Third Day, Sagredo congratulates Galileo for "the ease and clarity /whereby/ from a single simple postulate, he deduces the demonstrations of so many propositions."[56] As in Euclid's geometry, these propositions will become the path to many others, in part because of the preeminence of mechanics over the rest of physics.[57]

In a demonstrative science, the premisses must serve as warrant for the conclusions. How are the premisses of mechanics themselves warranted? Galileo sometimes suggests that they are "simple" or "natural" or "obvious," i.e. that they are self-warranting, once they are understood. In a reply to Antonio Rocco, an Aristotelian critic of the *Dialogue*, Galileo indicates how he arrived at one of his principles: "Proceeding discursively, I formulated an axiom which would not be doubted by anyone."[58] He goes on later to cite a "thought-experiment" (two bricks tied together falling at the same speed as one) of the sort that was standard in earlier natural philosophy. He is relying implicitly on a principle of sufficient reason in cases such as this, which are frequent in his work on hydrostatics and in his analysis of frames of reference in the *Dialogue*.

Axioms of this sort ought not, it would seem, require further confirmation. And Galileo often asserts that experimental confirmation is unnecessary for one or other of the key principles of his mechanics. He tells Rocco that he was "first convinced by reason rather than made certain by sense" when formulating the principle of the independence of weight and speed of fall.[59] In a letter to Ingoli in 1624, he asserted that before dropping a stone from the mast of a moving ship, "I had entirely convinced myself through reasoning that the result would be exactly what it turned out to be."[60] Sagredo agrees that the "natural light" of reason can determine without difficulty the truth of the crucial postulate of the Third Day: the velocity of a body on planes of different inclination depends only on the height of the plane.[61]

Salviati's response to this is interesting:

> You reason with good probability. But plausible considerations apart, I wish to increase the probability so much by an experiment that it will fall little short of necessary demonstration.[62]

It sounds as though intuitive plausibility alone is not enough. Experimental confirmation of the principle is also needed, just in case the intuition has been a faulty one. But if such confirmation is required, the resulting proof *cannot* be a necessary demonstration, as Salviati implicitly concedes. To say that it will fall only a "little short" of necessary demonstration is to say that

it is *not* necessary demonstration. Where strict necessity is concerned, no "falling short" is permissible. Even though his experiment "leaves no room for doubt about the truth of the assumption," a chastened Sagredo is much more cautious in his summing up:

The argument appears to me conclusive, and the experiment is so well adapted to verify the postulate that it may very well be worthy of being conceded as if it had been demonstrated.[63]

This is perceptive. A conclusive argument (as we saw above) is not yet a demonstrative one until the premisses are seen to be necessary in their own right. Galileo is trying hard to suggest that one can gradually come closer and closer to demonstrative argument by increasing the degree of experimental confirmation:

Hence let us take this for the present as a postulate, whose absolute truth will later be established for us by our seeing that other conclusions, built on this hypothesis do indeed correspond with and exactly conform to experience.[64]

This is no longer demonstrative science. Even if the "absolute truth" of the premiss could be established in this way, the reasoning is non-demonstrative. It is, in fact, hypothetico-deductive, i.e. the warrant for the premiss is not its own intrinsic self-evidence (as it would be were it demonstrative) but its confirmation by the verified inferences drawn from it. The warrant goes from conclusion back to premiss instead of in the direction of premiss to conclusion, as it does in demonstration. The reason to accept the premiss as true is that a verified proposition (an experimental fact, say) can be drawn from it. HD inference cannot conclude with certainty to the truth of a premiss or hypothesis unless a second argument be added which successfully eliminates all possible relevant hypotheses save one. When HD inference is causal in form, i.e. when the hypothesis introduces a cause, it is often called "retroductive" (Peirce), a term we shall need in Section 6 below. On the other hand, if an inference moves by way of generalization from a set of observations to an empirical law, it is called "inductive." Some would argue that this is a form of HD-inference since the inductive generalization, once formulated, is an hypothesis in need of further confirmation. We shall not discuss this issue here. Suffice to note that there is a relatively sharp distinction between retroductive and inductive inference. The former, for example, ordinarily introduces conceptual novelty, the latter does not. This distinction is of some importance in Section 5.

The Third Day focusses on uniformly accelerated motion, and derives a host of theorems, such as the squared-time law, from the original definition of the motion. Does this constitute a *demonstration* of the squared-time law? Galileo sums up the Third Day, as we have seen, by saying that he has demonstrated certain properties of motion in just the same way as Euclid demonstrated properties of circles. But this is not correct. And Galileo himself tells us why. Simplicio agrees that the various theorems about uniformly accelerated motion *do* follow from the definition. "But I am still doubtful whether this is the acceleration employed by nature in the motion of her heavy falling bodies."[65] An experiment is needed in order to show that this *is*, in fact, the acceleration actually found in nature.

Salviati responds:

> Like a true scientist, you make a very reasonable demand, for this is usual and necessary in those sciences which apply mathematical demonstrations to physical conclusions, as may be seen among writers on optics, astronomy, mechanics, music, and others who confirm their principles with sensory experiences that are the foundations of all the resulting structure.[66]

This reference to the *scientiae mediae* is helpful. We have seen that the demonstrative character of these sciences was supposed to derive entirely from their mathematical aspect. Insofar as they are applied to nature, their conclusions cease to be demonstrative properly speaking, unless the causal premisses are "seen" to be necessarily true in their own right. This is why they have to be confirmed by observations, which thus become (as Galileo correctly notes) the foundation or main warrant of the resultant theoretical structure. Mechanics taken as a branch of mathematics is demonstrative; mechanics taken as a part of physics is hypothetico-deductive.

It is tempting to present this insight as Galileo's considered view on the epistemic status of mechanics. Unfortunately, the passages just cited must be weighed against others where he still seems to be attributing full demonstrative status to mechanics. It will be helpful to focus on four different considerations that tended to conceal, perhaps even from Galileo himself, the non-demonstrative character of his mechanics considered as an account of nature, i.e. as a physics.

5. MECHANICS AS DEMONSTRATIVE IN PRINCIPLE

The first of these considerations has already been noted. Some of the principles of mechanics may easily seem so plausible as almost to take on the status of necessary truths. An assumption that nature is "simple" in its operation can readily transform a mathematical principle into a physical truth:

It is as though we have been led by the hand to the investigation of naturally accelerated motion by the consideration of Nature . . . /which/ habitually employs the first, simplest, and easiest means. . . . Thus when I consider that a /falling/ stone successively acquires new increments of speed, why should I not believe that those additions are made by the simplest and most evident rule? For if we look into this attentively, we can discover no simpler addition and increase than that which is added on always in the same way.[67]

The implication is that uniform acceleration is the "natural" sort of velocity-increment in the fall of heavy bodies.

Galileo clearly does not regard this as a coercive argument. In the paragraph just before it, he gives as his main reason for saying that the uniformly accelerated motion whose properties he is deducing is in fact the motion found in the world:

the very powerful reason that the essentials successively demonstrated by us correspond to . . . that which physical experiments show forth to the senses.

There are then, two different sorts of warrant for physical principles of this kind; one is the verification of observational inferences drawn from them, and the other is the internal simplicity (coherence, elegance) of the principles themselves. Galileo appears to be suggesting that both are important, and that they may complement one another, a theme of obvious relevance to later philosophy of science.

There is a second indication that Galileo leans rather more to the "demonstrative" side than the actual structure of his mechanics would seem to warrant. This is the frequent use of the Platonic-Archimedean theme of "impediments." One finds it in his earliest work; in an analysis of motion on the inclined plane in his *On Mechanics*, for example, he notes that he is "neglecting accidental impediments, which are not considered by the theoretician."[68] In the *Dialogue* he asks what happens when "in the ordinary course of nature a body with all external and accidental impediments removed travels along an inclined plane with greater and greater slowness according as the inclination is less."[69] Later, Simplicio and Salviati engage in a debate about the application of mathematics to physical reasoning. In Simplicio's view. "these mathematical subtleties do very well in the abstract but they do not work out when applied to sensible and physical matters." Mathematical spheres touch tangent planes at a point but "when it comes to matter, things happen otherwise."[70]

Salviati's response is to note an ambiguity in the (Platonic) notion of an "imperfection" of matter. If Simplicio means that matter is such that when a sphere is realized in it, it may touch a plane at more than one point, this is simply false. If he means that perfect spheres are never realized in nature, this could be true, but it would not follow from it that *if* such a sphere were to be realized in nature, it would not have the properties that geometry demands of it. It is difficult (in some cases perhaps impossible) to *realize* precise geometrical forms in matter. But to the extent they *are* realized, geometry can tell us their properties. Matter does not *alter* these properties; it only makes them difficult to reproduce exactly. Just as the businessman must allow for boxes and packings in computing the weights of his wares so the mathematical physicist (*filosofo geometra*):

> when he wants to recognize in the concrete the effects which he has proved in the abstract, must allow for the impediments of the matter, and if he is able to do so, I assure you that things are in no less agreement than are arithmetical computations. The errors lie, then, not in the abstractness or concreteness, not in geometry or physics as such, but in a calculator who does not know how to keep proper accounts. Hence if you had a perfect sphere and a perfect plane, even though they were material, you would have no doubt that they touched in only one point; on the other hand if it is impossible to have these, then it was quite beside the purpose to say that a bronze sphere does not touch /a plane/ at a point.[71]

An "impediment" is not, therefore, something which prevents, or lessens the force of, the application of mathematics to nature. Rather, it denotes a difficulty in *realizing* in the complexity of the material order the simple relations of the mathematical system. But to the extent they *are* realized, mathematics can tell us what to expect. This is clear enough. But in the *Discourses*, Galileo appears to forget this warning about the ambiguity of the Platonic notion of "defect." In the opening pages, where he is discussing the dependence of the properties of bodies on the size of the bodies, he alludes to "the imperfections of matter, which is subject to many variations and

defects" and is "capable of contaminating the purest mathematical demonstrations." So for the purposes of his discussion, he decides to abstract from all imperfections of matter "assuming it to be quite perfect and unalterable and free from all accidental change." By making it unalterable, he can construct "purely mathematical demonstrations no less rigorous than any others" of such properties as the resistance of bodies to breaking stress.[72] This is reminiscent of the *Timaeus*, where matter is presented as unstable, incapable of sustaining the intelligibility of form over time. Thus for mathematics to apply to nature, matter will have to be postulated to be unchanging.

In the Fourth Day, Simplicio raises an empiricist objection once again: "In my opinion it is impossible to remove the impediments of the medium" which makes it "highly improbable that anything demonstrated from such fickle assumptions can ever be verified in actual experiments."[73] Salviati's reply is very different from his earlier one; it sounds quite surprising from someone who had earlier written that the book of Nature is written in the language of mathematics:

> All the difficulties and objections you advance are so well founded that I deem it impossible to remove them. For my part, I grant them all, as I believe our Author would also concede them. I admit that *conclusions demonstrated in the abstract are altered in the concrete*, and are so falsified that horizontal /motion/ is not equable; nor does natural acceleration occur /exactly/ in the ratio assumed; nor is the path of the projectile parabolic, and so on. But on the other hand, I ask you not to reject in our Author what other great men have assumed, despite its falsity.[74]

This is precisely the understanding of "defect" that Galileo argued *against* in the *Dialogue*. Here the lack of fit between mathematics and physics is not seen to be due to the difficulty of precisely exemplifying mathematical forms in nature; rather, material nature is seen as not exactly following mathematizable norms, whether simple or complex. After an appeal to the authority of Archimedes, Galileo goes on to urge that in "practical operations" mere minutiae need not be taken into account; the mathematical science will give a reasonable degree of approximation, as it does in the case of architecture. Pragmatically speaking, then, we do not have to worry since the effects of the "impediments" are small.

Then again he changes course. If the mathematical assumptions *were* to be true of nature, he notes, "absolute proof" *would* be attainable. If we wish to allow for the degree of approximation involved in this, "we must remove from the demonstrated truth whatever is significant" in the approximation that has been made. The implication once more is that of the *Dialogue*: the "defect" is due to the difficulty in applying simple forms to a complex nature, and the degree of departure of nature om what the forms would lead one to expect is, in principle, calculable.

This is not, however, Salviati's last word on the topic. When the mathematical form used to understand material nature is not merely one of shape but of motion, the defect appears once more an irremediable one, due to the infinite variability of the matter itself:

A more considerable disturbance arises from the impediment of the medium; by reason of its multiple varieties, this is incapable of being subjected to firm rules, understood, and made into science.... No firm science can be given of such events of heaviness, speed, and shape, which are variable in infinitely many ways. Hence, to deal with such matters scientifically, it is necessary to abstract from them.[75]

This, of course, does *not* answer Simplicio's original objection to taking this *scientia media* as a demonstrative physics. He had urged that the "impediments" could not be removed, and Galileo is now agreeing with this but arguing that their effect is small so that an *approximate* account can still be given. An approximate physics is not, however, a demonstrative one. What is "demonstrative" here is the original mathematically-expressed scheme, and it is demonstrative only in the limited sense of being deductive in form. It is not demonstrative even in the sense in which geometry is, since its premisses are not asserted as *true*.

Galileo's apparent uncertainty in the face of this problem is crucial in discussions of his much-disputed Platonism.[76] Its importance for us here is to help us understand the ambiguous status that he is attributing to mathematics as an independent source of truth about nature. It is as though there is a mechanics which though absolutely "true" in its own right is not quite true of the physical world because of "impediments" which make the world an imperfect exemplification of the ideal realm. The role of experiment as confirmatory would in this event become equivocal; when a prediction is *not* confirmed, it could be blamed on "impediments." to "confirm" by means of experiments like that of the pendulum or the inclined plane is not (in *this* understanding of mathematical physics) to show in HD fashion the likely truth of the hypothesis that falling motion is uniformly accelerated. Rather, it is to show that the "true" mathematical theory of uniform acceleration

actually applies (more or less) to *this* material system. The outcome may seem the same. But the difference of emphasis is of crucial epistemic significance. In one case, experimental confirmation is integral to the status of mechanics as an HD science. In the other, it *is* a science (and a demonstrative one) because of its mathematical form, and observation only serves to determine to what extent it can count as a science of material nature.

The metaphor of "impediment" almost inevitably leads to this sort of ambiguity. Much safer is the notion of *idealization* which Galileo also uses. Here the arrow points in the other direction; one begins from the material system and "idealizes" in order to construct a simple theoretical account. The virtue of the idealization lies not in its own intrinsic formal properties but in the degree of understanding it affords of the complex physical world from which it takes its rise. Sagredo remarks: "You began by removing me somewhat from the sensible world, to show me the architecture with which it must have been built."[77]

Galileo "idealizes" in different ways. First, he simplifies the treatment of complex lines and shapes by taking them to approximate to a more easily treatable geometrical form. Second, he simplifies the *causal* factors, either by eliminating them, by holding them constant, or by gradually reducing them and inferring what would happen in the "limit" case. This last is elegantly illustrated in his handling in the *Discourses* of falling motion. He wants to know what would happen if moving bodies of different weights were dropped in a medium devoid of resistance. To find out, he tries a series of media of less and less resistance and finds that difference in weight affects speed less and less until in the most tenuous medium available, the difference is so small as to be almost unobservable. We may believe, he concludes, "by a highly probable guess that in the void all speeds would be entirely equal."[78]

This "asymptotic" method assumes that as the "impediments" are removed, nature is found to act in a simpler and simpler way, closer and closer to the mathematically-expressed law that is being proposed. The law is true of Nature, even though *in vacuo* fall does not occur in Nature. It is true as a limit, departures from which (due to causal "impediments") can later be calculated and allowed for. It is known to be true, not because of its geometrical simplicity, but because it is confirmed by a variety of experiments on pendulums and inclined planes. The spirit of this asymptotic approach is quite different from that of those remarks on "impediments" which would make them ultimately "incapable of being subjected to form rules."

In a letter to Baliani (1639), Galileo tries to explain what the methodology of the *Discourses* amounts to:

I argue *ex suppositione* about motion, so that even though the consequences should not correspond to the events of the natural motion of falling heavy bodies, it would little matter to me, just as it derogates nothing from the demonstrations of Archimedes that no moveable is found in nature that moves along spiral lines. But in this I have been, as I shall say, lucky: for the motion of heavy bodies and its events correspond punctually to the events demonstrated by me from the motion I defined.[79]

This is the Platonic Galileo once again; it would have "little mattered" to him (he says, only half-seriously, one supposes) if his account of motion had *not* been confirmed by the plane and pendulum. But if it had not been, what would the resultant "science" have been? Surely not a science of nature, a physics? No more than an elegant piece of deduction, the exploration of a mathematically-expressed supposition. The deep ambiguity in the conception of science underlying Galileo's mechanics is nowhere so evident as it is in this letter, one of the last he wrote on the topic of mechanics.

The use of the phrase "*ex suppositione*" above brings us to our third remark about Galileo's tendency to confer demonstrative status on his mechanics, despite the obvious tension between this and the method of confirmation-by-observation that he uses. William Wallace has recently argued that this phrase holds the key to Galileo's mechanics, and that its traditional scholastic sense allowed an *ex suppositione* reasoning to be strictly demonstrative in both the natural and the physico-mathematical sciences, leading thus to certain knowledge.[80] He concludes that this sort of reasoning:

was already at hand for Galileo . . . and it was capable of producing the scientific results he claimed to have achieved. Since the same cannot be said for hypothetico-deductive method in the modern mode, there is no reason to impose that methodology on "the father of modern science." Rather we should take Galileo at his word and see him neither as the Platonist nor as the hypothetico-deductivist he has so frequently been labelled, but as one who made his justly famous contribution in the Aristotelian-Archimedean context of demonstration *ex suppositione.*[81]

The difficulties with this claim ought to be plain by now. Wallace quotes both the Baliani letter and another to de Carcavi as two of the four instances of the *ex suppositione* phrase he has found in Galileo's work. de Carcavi relayed to Galileo Fermat's unhappiness with Galileo's assertion that a body falling from a tower on the rotating earth describes a semi-circular arc towards the center of the earth. Galileo notes in response that the correct curve is parabolic, and adds that his argument is *ex suppositione*, demonstrating conclusively certain properties of uniformly accelerated motion.

> I add further that if experience should show that such properties were found to be verified in the motion of heavy bodies descending naturally, we could without error affirm that this is the same motion I defined and supposed; and even if not, my demonstrations, founded on my supposition, lose nothing of their force and conclusiveness.[82]

Of course, they lose their claim to be *physics*. This is the same ambiguity once more. Galileo assumes that his "demonstrations" have the character of science independently of whether the definition from which they begin is true of the physical world. But this is only equivocally the case: as mathematical reasoning, yes, as physics no. The status of *ex suppositione* reasoning as *physics* depends critically on the sort of warrant the supposition is taken to have. There are only three possibilities. It may be intuitively self-evident. It may be an inductive generalization. Or it may be hypothetico-deductive, deriving its warrant from the observational verification of the inferences drawn from it. Wallace thinks that it is inductive, that it is found by repeated experiment "to be very nearly in agreement with what actually occurs in nature."[83] But even this is not enough to make it demonstrative in the Aristotelian sense, i.e. proposing a *necessary* connection of essence (or cause) and property (or effect), unless one takes "induction" (as Aristotle did) to be capable of yielding such intuitive self-evidence.

Galileo was no inductivist. True, he performed repeated experiments with inclined planes, charted the positions of the Medicean planets, tested the floatation behavior of bodies. He made use of rules of inductive method like those later popularized by Mill.[84] But he was not patient. He leaped quickly to generalization. Besides, he distrusted inductive methods as a source of science:

> I assure you that the movements, sizes, distances and arrangements of the orbs and stars will never be observed so accurately that they will not need endless corrections, even if the world were filled with Tycho Brahes or men a hundred times as good as he was. We can be certain that there are many movements, alterations, anomalies, and other things in the heavens as yet unknown or unobserved, and perhaps not even observable or explainable in themselves. Who can vouch that the movements of the planets are not incommensurable, and therefore susceptible to – or rather in need of – eternal emendation, since we can only deal with them as though they were commensurable?[85]

This somber assessment is, of course, at odds with other remarks that Galileo makes about the testimony of the senses, as well as about the language of the Book of Nature. It is reminiscent of Plato's remarks on observational astronomy. This recurrent Platonic distrust of the singular ought to be enough to make us wary of labelling him an inductivist, popular though that appellation once was in his regard.

Returning to Wallace's analysis, the instances of medieval *ex suppositione* reasoning that he thinks most appropriate to serve as illustration are the analyses of the lunar eclipse and the rainbow. Both of these phenomena are irregular, but their causes were thought to be understood. He goes on:

> *If* rainbows occur, they will be formed by rays of light being reflected and refracted in distinctive ways through spherical raindrops. . . . The reasoning, though phrased hypothetically, is nonetheless certain and apodictic; there is no question of probability or verisimilitude in an argument of this type. Such reasoning, of course, does not entail the conclusions that rainbows will always be formed. . . . But if rainbows *are* formed, they will be formed by light-rays passing through spherical droplets to the eye of an observer in a predetermined way, and there will be no escaping the causal necessity of the operation by which they are so produced. This process, then, will yield scientific knowledge of the rainbow, and indeed it is paradigmatic for the way in which the physical sciences attain truth and certitude in the contingent matters that are their proper subjects.[86]

One is reminded of Galileo's assurance in regard to the stone falling from the mast that it "*must* happen that way."[87] This long quotation is worthwhile because it illustrates so exactly the same ambivalence that underlay Galileo's own reasoning. Wallace and Galileo are both assured that the physical sciences attain truth and certitude and not just probability of verisimilitude. Wallace supposes that the theory of the rainbow is not hypothetical. But it *is* hypothetical: the notion of light-ray, the use of the spherical droplet model, the testing of the model by its correct prediction of the rainbow arc, are all characteristic of HD reasoning. There is no necessity about this analysis; a different one, using different notions of light and of raindrops, might also work. What *is* necessary is that *if* this supposition is made, certain inferences necessarily follow. The necessity is in the "following," not in the original premiss.

What further confuses the issue is the notion of contingency Wallace calls on. Rainbows and eclipses are "contingent" in the sense that they do not appear as part of the regular round. But they are just as natural, just as predictable as any other physical occurrence. The *ex suppositione* reasoning is used in their case not because they are "contingent," but because "impediments" of an unexpected sort *might* cause their non-appearance. This is a different point, one that would hold equally well for falling apples. Wallace is unwilling to allow that this sort of reasoning could be "hypothetical" because (like Galileo) he takes this term in a pejorative instrumentalist sense which would equivalently make Galileo's mechanics a fictive construction akin to Ptolemaic epicycles were its propriety to be conceded. But this is *not* the sense in which "hypothesis" has been used in recent centuries, nor is it the sense in which "hypothetico-deductive" is used today.

Our claim is, then, that *ex suppositione* argument *is* hypothetical, though medieval exponents of it were not always clear on this point (sometimes they were). And Galileo is *quite* clear in the letters to Baliani and de Carcavi that his mechanics *is* hypothetical (though he does not use the term), i.e. that the supposition may or may not apply to the physical situation he is trying to understand, and that the only way to discover whether it *does* apply is to test it by means of experimental confirmation.

A complication emerged above about the status of the law of falling bodies. Ought it be taken to be retroductive (i.e. involving a theoretical causal explanation)? This brings us to the fourth and final complication in our long attempt to decipher the epistemic status of Galilean mechanics. Galileo manages to dispense with causes. He does not *explain* motion; he merely tries to describe it. In modern terms, his mechanics is a kinematics only, not a dymanics. This was a serious shortcoming, one which would later be rectified by Newton who in the process proved that Galileo's "law" is in fact *not* a law but a rough approximation holding only for motions at the earth's surface.

Some recent writers have hailed Galileo's "elimination of causes in favor of laws"[88] as a major step to the modern ideal which they take to be nomothetic science. Clavelin, we have already discussed. Drake is almost as emphatic:

> There is perhaps something methodological new even in Galileo's ultimate presentation of his science of motion to his readers. . . . If that very form of presentation was not an open and explicit attack on the Aristotelian conception of science as an understanding of nature in terms of causes, it at least implied that everything of lasting value in physics could be presented in the form of precise laws, experimentally confirmed.[89]

If this is what it implied, it was, of course, wrong since theory has proved a far more potent instrument of understanding than has law. To state a law is merely to state an *explanandum*, something to be explained. Boyle's Law does not explain the behavior of gases; it merely describes it. Kinetic theory and molecular models have to be called on, if explanation is what is sought. Even in mechanics, where the concept of cause is least well understood and most controverted, Galileo's law cannot be said to *explain* the motion of falling bodies, whereas Newton's theory (in some sense) can. Modern science is causal through and through, though the structural causes postulated in chemistry, geology, biology, are remote from (and in some ways antithetical to) the sorts of causes described by Aristotle.[90]

Is it correct, however, to say that Galileo proposed to eliminate causal explanation, and that he thought his laws to be the ultimate in explanation in mechanics? We shall argue the falsity of this at more length in the next section where we shall be dealing with Galileo's extensive and creative use of causal explanation in cosmology. But even in mechanics, how did Galileo view his program? First of all, he is quite definite that the present is not "an opportune time to enter into an investigation of the cause of the acceleration of natural motion."[91] So far causal inquiry in mechanics has resulted only in "fantasies." Rather, he will limit himself to demonstrating "some attributes of a motion /uniformly/ accelerated (whatever be the cause of its acceleration)." And if the attributes so demonstrated are "verified in the motion of real bodies," the definition may be taken to apply to natural falling motion.[92]

There is no suggestion here that inquiry into the cause of motion is, in principle, unfruitful. What he seems to be saying is that first the proper *description* of motion must be obtained; a dynamics (we might say) must be preceded by a kinematics. In the *Dialogue*, this becomes clearer. In response to a question about the cause of the earth's motion Salviati responds:

> I did not say that the earth has neither an external nor an internal principle of moving circularly; I say I do not know which of the two it has. My not knowing this does not have the power to remove it. But if this author /Locher/ knows by which principle other world bodies are moved in rotation, as they certainly are moved, then I say that that which makes the earth move is a thing similar to whatever moves Mars and Jupiter, and which he believes also moves the stellar sphere. If he will advise me as to the motive power of one of these movable bodies, I promise I shall be able to tell him what makes the earth move. Moreover, I shall do the same if he can teach me what it is that moves earthly things downward.[93]

Now, while this *might* be read as an *ad hominem* argument or as a sceptical disclaimer of the chances of ever discovering these causes, it may also be read as a speculation regarding the unity of the causes of planetary motion. Drake, in fact, in a footnote to his translation of the *Dialogue*, took the last sentence (and the general context) to imply that Galileo may have "identified the cause of falling with the cause of planetary circularion"; it would "not be absurd," he concludes, to credit Galileo "with suspecting that a true comprehension of gravity would yield also an understanding of planetary motion."[94] Indeed, this does not seem in the least absurd. Galileo goes on to ridicule Simplicio's suggestion that "gravity" can explain it all; gravity is only a name, not an explanation, he retorts. But his own frequent use of the term, *gravità*, would indicate that he is using it causally. He often speaks of the "causes" of motion, as for example in the tidal argument of the Fourth Day. There is no sign anywhere in his work that he thinks these to lie permanently beyond understanding. What clearly hindered him in this regard was his distrust of "occult" explanations such as Kepler's notion of attraction. Newton's imagination, fortunately, was not thus bound.

Our concern here, however, is with the impact of the absence of causes on the epistemic status of his mechanics. It greatly simplified his task. He did not have to worry about causal hypotheses, and could thus plausibly suppose his science to be descriptive, at least in an approximate way. The notion of "fitting" principles to nature is thus far more straightforward than it would later be for Newton. The law of falling bodies can be verified *directly* (leaving aside the "impediments"). It can be inferred from the experiments (induction), at least in an approximate way. And the experiments can be inferred from it.[95]

This would *not* have been the case had the principle been a causal one. And it made the task of construing the new science as "demonstrative" both easier and harder. Easier because the troubles of retroduction to causes has been eliminated. Harder because "demonstrative" had traditionally meant causal inference in the context of natural philosophy. If, however, one accepts a weaker sense of the term as "necessary" or "concluding with certainty" (omitting all reference to causes), then one can make a plausible case for saying that the experiments have (inductively) confirmed the principles of the new mechanics, and that this guarantees the applicability of the Galilean definitions of motion to Nature, thus in turn guaranteeing the "demonstrative" character of the resultant science.

This may well have misled Galileo also into an overconfident estimate of the powers of geometry in formulating and extending the new science. Not only does he maintain that "all reasonings about mechanics have their foundations in geometry,"[96] He seems (as we have seen) quite often to *equate* mechanical proof with geometrical demonstration. What allowed him to do this was the almost accidental fact that he could formulate the law of falling bodies without involving either weight or resistance, the two non-geometrical factors that had figured so largely in earlier discussions of the law. If weight *had* made a difference to speed of fall, the pure geometrical approach would not have worked. Galileo says once that it was a matter of "luck" that his simple kinematic law applied to all bodies, whatever their material composition or their sizes.[97] In a sense this was true. But one might also add that there was an element of *bad* luck about this also, because the facility with which he was able to geometrize motion may have led him (as it certainly did others) to believe that the future of mechanics lay in this direction. Newton was to restore the balance, with his non-geometrical concepts of mass and force. But the will-o'-the-wisp of a "pure" geometry of motion, the ultimate in "demonstrative" science, is still occasionally to be seen dancing through the speculative fields of modern physics.[98]

6. THE RETRODUCTIVE SCIENCE OF COSMOLOGY

The demonstrative ideal of science, as we have seen, could work only in mathematics or where physical premisses could be plausibly regarded as intuitively evident. For the latter to be the case, the natures involved had to be close at hand, familiar. Galileo's telescope opened up a new and puzzling realm. No longer immutable and circularly moving entities, sharing a single "fifth" essence, the heavenly bodies now demanded a new kind of science. Since they lay beyond the scope of the direct observation on which Aristotelian demonstrative physics had rested, a new and less direct mode of proof would have to be worked out. Descartes was more concerned with the motions of the heavenly bodies than with the strange events the telescope had disclosed, and so he could still hope to encompass the stars in his grandiose geometrical mechanics. For him, the multiplicity of the causal lines only became a problem when the scientists' gaze descended from the heavens to the processes of earth. But Galileo, when faced with such problems as the nature of comets, could only shake his head and remind his hearer of "nature's bounty in her variety of methods for producing effects."[99] As he saw, there were no resources within the compass of demonstrative science for handling such questions.

There was a second, even less promising, context where demonstrative science could clearly find no purchase. This was in the realm of the very small, what Newton later called "the invisible realm." In the Aristotelian tradition, causal explanation in terms of atoms and the like had been excluded, for reasons that were quite central both to Aristotle's theory of knowledge and to his teleology. But with the growth of atomism in the sixteenth century, the possibilities of such explanation were once again canvassed.[100] The trouble was that it seemed almost impossible to get such explanation going, even in the weakest hypothetical form. Indeed Locke would later give strong arguments to show that a science of the atoms, an explanation of the observed properties of things in terms of the minute corpuscles composing them, could never be constructed. Galileo had not heard these arguments, and he talked freely of atoms:

> The tiny descending particles are received upon the upper surface of the tongue, penetrating and mixing with its moisture; and their substance gives rise to tastes, sweet or unsavory accordingly as the shapes of these particles differ, as they are few or many, and as they are fast or slow.[101]

Later he goes on to explain the operation of fire by means of the "number and velocity of the fire-corpuscles," and adds that "in addition to shape, number, motion, penetration, and touch," there is no further quality in them called "heat."[102] Only a few pages before he had castigated Grassi for his use of "probable arguments, conjectures, examples, analogies and other sophisms," and contrasted with this his own commitment:

> to the rigor of geometrical demonstration. . . . For just as there is no middle ground between truth and falsity in physical things, so in rigorous proofs one must either establish his point beyond any doubt or else beg the question inexcusably.[103]

Surely Galileo did not think that geometrical demonstration could extend downward to the fire-corpuscles? It is strange that he never directly addresses this issue. Like Descartes, he apparently felt free to invent plausible hypotheses in this domain without holding himself to any real confirmation of an observational sort. Galileo later asserted for example that a "fume" is given off when the "pores" of glass are opened. When Grassi objected that it is not in fact observed (as it should be) when the glass is broken, Galileo undisturbed responded: "When the plate is broken in two, a fume or exhalation rises, but it remains invisible, because it does not carry with it the light dust by which it becomes visible."[104] In the *Discourses*, he prefaces a passage about fire-corpuscles with the disclaimer: "I do this not as the true solution but rather as a kind of fantasy full of undigested things."[105] But on the very same page, he proposes to explain the coherence of bodies by the same "force of the void" that keeps two slabs of marble from separating, now operating as "tiny voids on the most minute particles." And since "for any effect there is one unique and true and most potent cause, if I can find no other glue, why should I not try to see whether this cause, the void, already found, may suffice?" He seems to be serious about the possibility of causal explanation here. And to make it demonstrative (or at least conclusive), he will assume a unique effect-to-cause inference, eliminating the need for retroduction.

Galileo's discussions of the "invisible realm" are two few and too casual to build any kind of case on them. The matter is quite otherwise when we come to his cosmology. Then his reasoning becomes explicitly (and often elegantly) retroductive. Out of the many examples one could choose, three stand out: his discussions of the lunar surface, of sunspots, and of comets. The first of these is a particularly well-worked-out instance of analogical reasoning, part of it explicitly retroductive in form.

The problem with causal reasoning in cosmology, as Galileo realizes, is the remoteness and possible unfamiliarity of the causes; the problem with retroductive inference to such causes in that it is very difficult to test, since direct experiments are not possible. "My conclusion," Galileo remarks, "is that it is indeed possible to discover some things that do not and cannot exist on the moon, but none which I believe can be and are there, except very generally."[106] Thus he will not speculate on the kinds of organisms that might be found there. But the *general* features of the lunar surface appear to resemble those of earth; he warns however that "the lunar globe is very different from the terrestrial although in some points conformity is to be seen."[107] His method in the *Starry Messenger* and in the *Dialogue* can be represented in two rather different ways. He looks for likely points of resemblance, and checks to see whether they are found (method of analogy). He tries to interpret observed lunar phenomena by supposing them to be explained by a known causal process similar to one obtaining on earth (retroduction). And he *tests* (or at least claims to test) his hypothesis that the gross features of the lunar surface are similar to those of earth by inferring how the *earth* would look from a distance, and then checking to see whether something like this is found in the case of the moon (retroductive or HD testing).

In the *Dialogue* he lists seven points of resemblance: spherical shape, opacity and non-luminosity, broken solid surface with mountains and plains, contrast between darker and lighter areas which on earth correspond to sea and land but on the moon could be very flat plains and broken terrain, phases (as seen from the other) of precisely the same period, moonlight and earth-shine as night illumination, the capacity to be eclipsed by the other. All of these have to do, one way or the other, with *light* and his main assumption is that light will behave in the same way on the moon as it does on earth. The methodological status of the resemblances he draws differs from one instance to the other, we shall see.

The spherical character of the moon is easiest to show, and is in fact "indubitable." The moon is seen by us exactly as a sphere illuminated by the sun would be seen. He is brief on this, since it was a familiar argument going back to Greek times. The phases of thc earth as seen from the moon are a purely theoretical consequence, a resemblance that *would* be observed were we in a position to do so. On the other hand, we *do* observe that the line dividing the illuminated part of the moon from the dark part is broken as though by mountains. There are patches of light in the dark part near the edge that look as mountain peaks illuminated by the sun would look while the lowlands are in shadow.[108] And these patches grow before sunrise or shrink after sunset, remaining consistently in the same places day after day, just as they would were mountains to be the causes. This sort of causal action is already well-known to us. Why, then, is the outer rim of the moon, as we see it, not jagged, since the mountains of the moon (if they are so interpreted) are proportionately higher than those of earth, some as high as four miles (Galileo gives an ingenious, and quite correct, way of estimating this)? He answers that mountain ranges tend from a distance to take on a more or less flat profile, because of the averaging-out of peaks at different distances.

Most of his discussion centers around the appearances of darker and lighter patches on the moon. By appealing to some simple "experiments" with plane and spherical mirrors, he is able to show that the moon cannot be thought of as a mirror; rather, it is broken surface such as that of earth that at a distance appears bright, by scattered light. Sagredo says enthusiastically: "If I were on the moon itself I do not believe that I could touch the roughness of its surface with my hand more definitely than I now perceive it by understanding your argument."[109] Salviati is more cautious: "do not attribute more to my demonstration than belongs to it." It is still not clear just how the moon *would* look if it were mirror-like; a test with a spherical mirror shows a blinding reflection – and yet the general illumination on a wall is not increased by the mirror. He makes his point: one must proceed very cautiously "in giving assent to what is shown by argument alone. There is no doubt that what you say is plausible enough, and yet you can see that sensible experience refutes it."[110] His solution of this particular difficulty is to introduce (as he did in the debate with Grassi on comets) an "irradiation" in the eyes which distorts visual estimates of light intensity. The anomalous experimental result is thus explained away, and the original hypothesis (that the reflecting surface of the moon *is* like that of earth and is *not* like a mirror) is sustained.

The most striking inference that this hypothesis yields is that the earth would be seen from the moon as a shining body just as the moon is seen on earth. And its light would be even stronger. It follows that the part of the moon not illuminated by the sun will be lit up by "earthlight" to an extent which depends on the relative configuration of sun, earth and moon. Galileo notes that this faint illumination of the "dark" part of the moon is already known, and that its observed intensity-changes verify the terrestrial origin of the light responsible for it. Here then is confirmation of the original hypothesis about the nature of the lunar surface.

Galileo has Simplicio formulate an alternative hypothesis more in keeping with Aristotelian principles in a rival effort to explain the varied phenomena. He has no difficulty in showing that Simplicio's hypothesis is too general; it cannot account for even *one* of the specific results that Galileo's own hypothesis so elegantly explains.[111] The discussion ends with yet another warning from Galileo that the contrast between dark and light on the moon (which on earth would be due to the contrast of sea and land) may be due to any one of a number of other causes: The darker patches could be forest or soil of a darker color, for example. These are "other ways in which the same effect could be produced," and there may well be "others that we do not know of."[112] Thus, the one-effect-one-cause principle he elsewhere relies on is evidently not universally applicable. But he seems quite sure that the general analogy he is drawing is borne out by such a wide variety of verified consequences, and depends on properties of light that are so well attested to, that it can be called conclusive.

In his long debate with Scheiner about the nature of the sunspots, he adopted a similar approach.[113] Scheiner believed the sunspots to be dark bodies, like planets, orbiting the sun; Galileo argued that the sunspots were on, or very near, the solar surface, and that the nearest analogy to them in terrestrial terms is that of clouds, "which are produced and dissolved in brief times, endure for longer or shorter periods, expand and contract, easily change shape, and are more dense and opaque in some places than in others."[114] Yet he is not to be thought to assert that they *are* clouds: "I merely say that we have no knowledge of anything that more closely resembles them."[115] This part of his account is to be taken as hypothetical, nothing more. Also, their rotation *could* be explained by a rotation either of the sun or of a surrounding medium, "but to me it seems much more probable that the movement is of the solar globe," and he goes on to give reasons for this.[116]

What he is *sure* of is that a geometrical analysis of the apparent motions, separations, and changes of shape of the spots puts them on or close to the sun's surface. This is a matter of deduction, of rigorous proof, of demonstration; he uses all the familiar phrases. Yet as he phrases the argument, it comes across as HD in form: all the appearances "agree with the hypothesis" that the spots are contiguous to the surface.[117] The phenomena "correspond exactly to what *ought* to appear if the spots *are* contiguous to the surface."[118] The alternative hypothesis which is "contradicted" by the phenomena is faced with obvious inconsistencies and contradictions,[119] and so his own hypothesis, being the *only* one that can fit, is proved.

The care and caution characteristic of these two discussions is, alas, almost entirely absent from his controversy with Grassi over the nature of comets. Whether it was the passion of hurt pride, the pressure of illness, the worry induced by a deepening conflict with Roman authority, it is in any event hard to see in the author of the *Assayer* the same man who penned the *Dialogue* or the *Starry Messenger*. The argument is slipshod. The language is intemperate, even for the polemic of that day. His unfortunate opponent, whom he characterizes as "a blind chicken poking its beak into the ground, now here and now there, hoping to find some grain of wheat to bite and peck at,"[120] is repeatedly misinterpreted. It is perhaps not *too* far-fetched to attribute some of the rancor of this demeaning controversy to the fact that both men were uneasy with hypothesis. Galileo, in particular, keeps attributing the status of geometrical demonstration to his arguments, and waxes sarcastic about the "sophisms" of his opponent. The original treatise (which he keeps insisting Guiducci wrote, though we now know that this is not true) is, he says, a demonstration "purely geometrical, complete, and logical."[121]

In his discussions of the lunar surface and of sunspots, Galileo had made good use of geometrical optics. But in the controversy with Grassi, this resource failed him. It failed him, first, because his view of the nature of comets (that they are refractions of sunlight through vapors ascending from the earth) would have required in its defence a much greater knowledge of optics than he possessed, and second, because it was quite manifestly at variance with the known facts of cometary orbits. The original *Discourse* is mainly a critique of the view held by Tycho and Grassi that comets are "temporary planets." Two of the reasons Galileo urges most strongly against this view are that it attributes non-circular orbits to comets (he castigates this abandonment of circularity as an "extravagance," a "fancy"[122]), and second, that it implies immense orbits: "how many worlds and how many universes to give it space enough for an entire revolution."[123] One recalls this precise objection later being placed in the mouth of Simplicio!

In the original *Discourse*, Galileo's own theory is put forward almost casually. No precise verifiable predictions are drawn from it; in fact, scarcely any argument for it is given. Much more seriously, there is an obvious objection against it; his theory would imply that comets always move toward the zenith, and that their path could never subtend an angle greater than 90°. Not only was this known to be false, but Galileo openly admitted this:

> I shall not pretend here not to know that if the material in which the comet takes form has only a movement straight and perpendicular to the earth's surface (that is, from the center toward the sky), the comet should appear to be directed exactly toward the zenith; yet it did not appear so, but declined toward the north. This forces us either to change what has been said, or else to retain it while adding some other cause for this apparent deviation. I cannot do the one, nor should I like to do the other.[124]

He goes on to say that for a sure determination of these matters, a firm knowledge of the order of the universe would be needed, but "in our age we still lack this; we must be content with what little we may conjecture here among shadows, until there shall be given to us the true constitution of the parts of the universe." This from someone who only a few pages earlier had assailed

Tycho for his "fantasies"!

Grassi's response in the *Libra* was not so much to argue for the Tychonic view as to urge a whole series of negative HD arguments (falsified inferences from Galileo's hypothesis), and to criticize the arguments Galileo put forward against Tycho. Unfortunately, he repeated his erroneous argument for the great distance of comets from the earth which was based on the assumption that the magnification of the telescope depends upon the distance of the object. In the *Assayer*, Galileo has no difficulty in demolishing this, though he never manages to account for the main evidence in its support (the fact that the stars are not magnified), and he is unable to separate the entangled issues of degree of magnification, clarity of image, brightness of image, and "irradiation" due to the eye. He emphasized this last, even though he could give no good reason why it would not affect telescopic observation as much as any other (and his argument depended crucially on this).

This controversy affords an example of retroductive argument badly handled. Not only does Galileo not offer confirmation of any sort for his main hypothesis, but disconfirming evidence is ignored and belittled. Nonetheless, one notes the dialectical skill Galileo showed (and which Grassi did not altogether lack) of attacking an opponent's argument by deriving from it a variety of testable conclusions and then focussing on those that could be falsified by observation. We find Galileo's initial protestations of the modesty of his own claim disingenuous, to say the least: "I frankly confess that I am vanquished and almost totally blind when it comes to penetrating the secrets of Nature."[125] Yet it is clear that the methodological tangles of this controversy, perhaps more than any other in which he engaged, must have impressed upon him the limitations of the demonstrative ideal of science he still strove to uphold.

There was one other context, however, where these limitations weighed on him even more, though he was much less willing in this context to admit the propriety of acknowledging these limitations than he had been when discussing sunspots and comets. To this context we finally turn.

7. THE COPERNICAN DILEMMA

It was in the Copernican controversy that Galileo's troubles with demonstrative science came to a head. There were two quite separate reasons for this.

First, in order to *demonstrate* the movement of the earth, he would have to use causal argument. Yet he had deliberately laid aside causes in his own kinematic treatment of motion. He introduces the ingenious analyses of the Second Day of the *Dialogue* in this way:

> The true method of investigating whether any motion can be attributed to the earth, and if so what it may be, is to observe and consider whether bodies separated from the earth exhibit some appearance of motion which belongs equally to all.[126]

And he goes on to give a series of seven "reductive" arguments, showing how much *simpler* it is to postulate the earth's rotation than to suppose the universe of stars to be in steady circular motion around the earth.[127] But arguments from simplicity, he notes, do not *exclude* the possibility of the more complicated alternative; "I do not pretend to draw a necessary proof from /them/, merely a greater probability."[128] They are not "inviolable laws," only "plausible reasons," which "a single experiment or conclusive proof to the contrary would suffice to overthrow."[129] These reductive arguments are HD in form, but their dependence on "simplicity" criteria makes them rather different from the more usual retroductive arguments which rely on confirmation through testing of predictions.

Further, his analysis of shared motion had seemed to show that the motions of the earth could have no perceptible effects at the earth's surface. So where could one look for demonstrative proof? In the Fourth Day, he "plunges" on a causal argument from the tides that he hopes will be demonstrative. But in the Second Day, he had attributed a circular inertia to unattached bodies on the earth's surface, so that no *differential* (and, thus perceptible) effects of the earth's motions could be admitted in regard to such things as water or air. There is thus no way to make the Second and Fourth Days consistent with one another. And although the motions of the earth *do* affect the waters of earth, they do not (even if the inertial notions of the Second Day be replaced by those of the *Discourses*) explain the tides.

There was one further move that Galileo had tried with success in his other HD arguments in the effort to make them conclude with certainty, the move that both Kepler and Descartes recommended in order to get "true knowledge." That was to exclude all hypotheses save one. Much of the argument of the *Dialogue* is directed against the physics of Aristotle and the astronomy of Ptolemy. The other "Chief World System" of his title is an amalgam of the two, and clearly he *had* succeeded in undermining Aristotle and refuting Ptolemy. The suggestion not only of the title he chose for the *Dialogue* but of the entire development of its argument was that with the elimination of one of the systems, the other could be taken to be established. The Tychonic third alternative is simply ignored, despite the fact that many of the most influential astronomers of the day subscribed to it. The reductive arguments might work against the strict Tychonic scheme. But this was not to *demonstrate* the truth of Copernicanism. Only a proper *causal* argument could do that, as Galileo well knew, and none was in sight.

Why could Galileo not have settled for a weaker claim, namely a high degree of plausibility for the Copernican theory? Given his views of science, one can see why. It would have been equivalent to saying that science could not really separate the claims of Copernicus and Tycho. And Galileo's intuitions as a scientist rebelled against this. But his intuitions in regard to the validity of Copernicianism were formed, in point of fact, not by any demonstrations he constructed, but by the weaker reductive arguments which a (much) later generation of scientists would be perfectly happy to accept as legitimate and convincing.

There was, however, a second and quite different reason why the notion of "demonstration" came to plague Galileo in this controversy. In the *Letter to the Grand Duchess Christina*, he hesitated between two views of the relation between the Bible and natural science.[130] One series of arguments he gives leads quite cogently to the conclusion that the language of the Bible is necessarily the common language of the time each book was composed. Thus, such turns of phrase as the "standing still" of the sun are no more than turns of phrase, idioms of no special scientific significance. The implication of this would be that the Bible and physical science have no mutual significance and cannot therefore, conflict.

The *Letter* contains, however, a quite different and incompatible view, the traditional Augustinian one according to which the scientist must provide a "conclusive demonstration" of the truth of his claim before the theologian may be called on to give a non-literal interpretation of the apparently conflicting passage. The reason is (in Galileo's own words, modelled on those of Augustine) that the authority of God "ought to be preferred over that of all human writings which are supported only by bare assertions or probable arguments, and not set forth in a demonstrative way."[131] The merest possibility that a literal interpretation should be the correct one counts for more than any reasoning short of full and proper scientific demonstration.

Thus when he is discussing what status a scientific theory such as that of Copernicus should have in order to warrant a metaphorical interpretation of Biblical phrases indicating a moving sun, he uses terms like "manifest experience and necessary proof," "physical certainty," "rigorous demonstration" dozens of times. In this way he commits himself to a *demonstration* of the earth's motion on theological grounds, the importance of which the 1616 decree of the Congregation of the Index had assuredly brought home to him.

It seems more likely that the first of the two hermeneutic theories above would have been his own preference. Yet he could not forget that his critics, operating quite justifiably with the traditional Augustinian theory of Scripture would insist on demonstrative proof of the earth's motion as the only legitimate means of setting aside the literal reading of the controverted phrases from the Bible. Admittedly, the decree of 1616 had proscribed the defence of Copernicanism, and Pope Urban had emphasized this once more in charging Galileo to treat Copernicanism in the *Dialogue* as a "hypothesis," i.e. as a useful mathematical fiction, in the traditional understanding of this term in mathematical astronomy.

So Galileo was caught. He could not, as a scientist, accept the implications of Urban's use of "hypothesis" for the Copernican view. Yet in order to argue against it, it would not suffice to urge Copernicanism as a good "hypothesis" in Galileo's *own* sense of that term (i.e. as a plausible and even perhaps the best-warranted account). What was needed was a fully *demonstrative* account in order to overcome the Augustinian objection he had himself acknowledged. Were he to succeed in this, then the apparent conflict between astronomy and the Biblical phrases would be resolved. And if enough people of goodwill could see this, perhaps the decree of 1616 would be rescinded, as the grounds for it would be undermined.

So Galileo's reasoning may have gone during those tense years in which he was trying to shape the proper proof-structure for the *Dialogue*. We marvel today at its ambiguities, and wonder why he did not (as a modern scientist would) content himself with a weaker claim than demonstration. Why did he push, against all odds scientific and ecclesiastical, to make the *Dialogue* into a conclusive demonstration? We have encountered two possible answers. One lies in his inbred disdain, as a scientist in the "old" tradition, for anything less than certainty in a scientific claim. The other lies in the peculiar cast given the discussion by the Augustinian view of knowledge.

The frequent use of anamnesis in the *Dialogue* is assuredly connected with this imperative to make a demonstrative case of a non-demonstrative argument. At one point Sagredo describes various possible behaviors of rolling hoops which Simplicio finds unbelievable. Sagredo says:

> I say to you that if one does not know the truth by himself, it is impossible for anyone to make him know it. I can indeed point out *things* to you, things being neither true or false. But as for the true, that is the necessary, that which cannot possibly be otherwise, every man of ordinary intelligence either knows this by himself or it is impossible for him ever to know it. . . . Therefore I tell you that the causes in the present problem are known to you, but are perhaps not recognized as such.[132]

If science is of the necessary, one can see that *only* something like anamnesis would serve to communicate it. Sagredo goes on to push Simplicio in regard to the effects of rolling and spinning, and leads him to admit the desired principle on the basis of observational knowledge he already possessed, but whose implications he had not explored. Feyerabend argues that Galileo adopts the device of anamnesis as a way of concealing what he is *really* doing which is to revise Simplicio's (and thus the reader's) experience. It is, he says, a clever tactical move, a ruse.[133]

If our argument above is correct, however, argument by anamnesis is, if not *required*, at least a highly appropriate means of revealing necessary connections between concepts rooted in experience, if the claim to necessary knowledge in the classic sense Galileo never really despaired of is to be sustained. Anamnesis is directed to someone who in Feyerabend's words "has developed the art of using different notions on different occasions without running into a contradiction."[134] Its aim is to force a synthesis of elements, a resolution of tensions. And in Galileo's case, there was the hope that the "truth" thus arrived at, without need for the routine of hypothesis and confirmation, could furnish the basis for a respectable demonstrative science.

8. CONCLUSION

We have traced two very different conceptions of science animating Galileo's work. One, the demonstrative ideal, he inherited from the Greek tradition and never abandoned, even though it led him into the gravest difficulties, especially in his cosmology. This is the conception of science he formally espoused throughout his career. The other is the retroductive notion of science which is exemplified especially in his discussions of phenomena whose causes are either remote (comets, sunspots), enigmatic (the motions of the earth), or invisible (atoms, the force of the void).

He used retroductive inference with great skill, and tried where possible to make it conclusive by eliminating all hypotheses save one. One could argue[135] that all his science, even his mechanics, is ultimately hypothetico-deductive in form, even though it *attempts* to be demonstrative. Yet to put it this way would not do justice to his stubborn refusal to consider anything less than rigorous demonstration as proper "science." We can hail Galileo and Descartes as the pioneers in the development of a new conception of science that would ultimately replace the older demonstrative one. But if we do, we ought to recognize how reluctant Galileo would have been to accept this honor.

University of Notre Dame

NOTES

* This paper is a greatly enlarged version of an earlier one 'The Conception of Science Underlying Galileo's *Discorsi*,' delivered at the XIII International Congress of the History of Science, Moscow, 1971, (*Proceedings* 5, 250–8), and later under the title 'Two Conceptions of Science in Galileo's Work,' at Yale University, Indiana University, and the Hebrew University of Jerusalem. I am grateful for the many suggestions I received on those occasions. The main work of preparation of the paper was done during a period of research in 1973–74 supported by the National Science Foundation, whose generous aid I would wish to acknowledge. Like all Galileo scholars, I owe a particular debt of gratitude to Stillman Drake whose herculean labors have made our task so much easier than without him it would have been. This essay owes much besides to the lively discussions we have had over many years.

1 See, for example, A. C. Crombie, 'Galilée devant les critiques de la posterité,' Conference du Palais de la Découverte, Paris, 1956, Section 3.

2 *Against Method*, London, 1975, p. 160; 'Problems of Empiricism II', *The Nature and Function of Scientific Theories, (ed. by R. Colodny), Pittsburgh* 1970, p. 323.

3 *Against Method*, pp. 89, 84, 81.

4 *Op. cit.* p. 89.

5 Preface to the second edition, N. K. Smith translation (London, 1956), p. 14.

6 Kant, *loc. cit.*

7 *Op. cit.* p. 154.

8 *Philosophy of the Inductive Sciences*, London 1947, vol. 2, p. 220.

[9] *The Science of Mechanics*, Chicago 1893, p. 140.
[10] See Maurice Clavelin, *The Natural Philosophy of Galileo*, Cambridge (Mass.), 1974; Dudley Shapere, *Galileo: A Philosophical Study*, Chicago 1974.
[11] Book 3, chapter 3–5.
[12] *Physics*, II, 2, 194a 7.
[13] *Posterior Analytics*, I, 9.
[14] I have discussed two of the leading figures in this 'mathematical' tradition, in 'Ptolemy on Saving the Phenomena', prepared for the XV Intern. Congress for the History of Science, Edinburgh 1977, and 'Mathematics and Physics in the Work of Alhazen', Conference on the relations of history and philosophy of science, Jyväskylä (Finland), 1973, Proceedings to appear in *Boston Studies in the Philosophy of Science.*
[15] See J. Gagné, 'Du *quadrivium* aux *scientiae mediae*', *Actes Congr. Intern. de Philosophie Medievale*, Montreal 1969, 975–86.
[16] This is a recurrent theme in Pierre Duhem's great work, *Le système du monde* (Paris, 1913–17, 10 vols.)
[17] *Posterior Analytics*, II, 19.
[18] *Op. cit.* (translated by H. Tredennick), London 1960, 89b, 15–21.
[19] Though it will not be known as cause, demonstration of the *oti* may show the nearness of planets by making use of known facts about the steady light of planets, and the correlation between distance and twinkling. But only if this correlation is known to be a *causal* one, i.e. that distance *causes* twinkling (so that non-twinkling objects must be reckoned to be near), can the nearness of thc planets be not only *established* but reckoned a *cause.*
[20] *Op. cit.* 79a, 24.
[21] This is done very well by Nicholas Jardine, 'Galileo's Road to Truth and the Demonstrative Regress', *Studies Hist. Philos Science* 7 (1976), 277–318. He argues that the Paduan *regressus* was not the foreshadowing of experimental and hypothetical method that Randall claimed it to be, and that in its manner of going from effect to proximate cause and back to effect, it left unresolved the main difficulties in the Aristotelian scheme. He also claims that far from being influenced by this notion of *regressus*, Galileo rejected it explicitly. This last is too strong a conclusion from the texts he cites, but what *is* shown is that the analytic procedure described by Galileo in his *mechanics* is of a different sort (back to axiom rather than to proximate cause), and further, that there is not much evidence of direct influence on his own work of the specifically Paduan version of the *oti-dioti* problem. We shall show below that in contexts other than the mechanics, however, Galileo relates demonstration to *causes* rather than to *axioms.*
[22] As a young man, he wrote a detailed commentary on the *Posterior Analytics*, the *Disputationes de Praecognitione et Demonstratione*, in which he followed the '*oti-dioti*' distinction by reducing scientific demonstration to the usual two kinds, *propter quid* and *quia*, insisted that *ex suppositione* argument could not be regarded as truly scientific, and argued that *regressus* is needed in physics (unlike mathematics) because physical causes may be less known than their effects, so that one has to work back to cause and then down to effect again to achieve complete demonstration. This work is not included in Favaro's National Edition of Galileo's works, and its exact status is still under discussion. See A. C. Crombie, 'Sources of Galileo's Early Natural Philosophy', in *Reason, Experiment and Mysticism* (ed. by M. L. Righini Bonelli and W. R. Shea), New York, 1975, pp. 157–75.

[23] *Opere, 8,* 190. The Drake translation (Madison, Wis., 1974) will be used; separate page-references to it are unnecessary since it provides the pagination of the National Edition.
[24] This is well brought out by E. W. Strong in the chapter on Galileo in his *Procedures and Metaphysics*, Berkeley 1936.
[25] *Dialogue*, p. 50; *Opere, 7,* 75. The translation used will be that of Drake (Berkeley, 1953), slightly modified.
[26] *Dialogue*, pp. 50, 328; *Opere, 7,* 75, 355.
[27] *Dialogue*, pp. 53–4; *Opere, 7,* 78.
[28] *Dialogue*, p. 406; *Opere, 7,* 432.
[29] Notably in the *Letter to the Grand Duchess Christina*, where it (or a near equivalent) appears on almost every page.
[30] As Crombie reports the contents of this (as yet unpublished) commentary, several odd features emerge (*op. cit.*, pp. 172–3). True knowledge can only be of real beings, unlike the entities of mathematics. Physical sciences which draw upon mathematical principles (the *scientiae mediae*) do not generate truly scientific demonstration on their own account. It would seem to follow therefore that neither mathematics nor mechanics can of themselves give scientific knowledge. Further, *demonstratio quia* (*oti*) is said to demonstrate the existence of an effect, and to begin from true and necessary premisses; as we have seen, this was not Aristotle's view.
[31] *Opere, 6,* 67.
[32] *On Mechanics*, Drake translation (madison 1960), p. 151; *Opere, 2,* 159.
[33] *Op. cit.*, p. 157; *Opere, 2,* 165.
[34] *Discourses, Opere, 8,* 127.
[35] *Dialogue*, p. 356; *Opere, 7,* 383.
[36] *Loc. cit.*
[37] *Dialogue*, p. 103; *Opere, 7,* 128–9.
[38] *The Assayer*, Drake translation, p. 236; *Opere, 6,* 281.
[39] *Discourse on Method* (translated by P. J. Olscamp), Indianapolis 1965, Part 6, p. 52.
[40] *Dialogue*, p. 101; *Opere, 7,* 127; see also *Dialogue*, pp. 39–40; *Opere, 7,* 64.
[41] *Loc. cit.*
[42] M. Clavelin, *The Natural Philosophy of Galileo*, Cambridge (Mass.) 1974, p. 390.
[43] *Op. cit.*, pp. 395, 409.
[44] *Op. cit.*, p. 411.
[45] *Op. cit.*, p. 407.
[46] *Op. cit.*, p. 411.
[47] *Op. cit.*, p. 392.
[48] *Discourse on Bodies in Water*, Salusbury translation, p. 18; *Opere, 4,* 79.
[49] *Dialogue*, p. 417; *Opere, 7,* 443.
[50] *Discourse, Opere, 8,* 296.
[51] *Dialogue*, p. 424; *Opere, 7,* 450.
[52] *Dialogue*, p. 460; *Opere, 7,* 484.
[53] *Dialogue*, p. 445; *Opere, 7,* 471.
[54] *Discourses, Opere, 8,* 296.
[55] *Op. cit.*, p. 54.
[56] *Op. cit.*, p. 266.
[57] *Op. cit.*, p. 267.

[58] Reply to A. Rocco's *Esercitazioni Filosofiche, Opere, 7,* 744. See also *Discourses, Opere, 8,* 108–9.
[59] *Op. cit.*, p. 731. See W. R. Shea, *Galileo's Intellectual Revolution*, London, 1972, p. 157.
[60] *Opere, 6,* p. 545. Shea, *loc. cit.* It is interesting that in the *Dialogue*, he strengthens this by omitting any reference to his having actually performed the experiment. It is as though he wants to underline that for Salviati an experiment would be altogether redundant as confirmation in this case (p. 145; *Opere, 7,* 171).
[61] *Opere, 8,* 205.
[62] *Loc. cit.*
[63] *Opere, 8,* 207.
[64] *Opere, 8,* 208.
[65] *Opere, 8,* 212.
[66] *Loc. cit.*
[67] *Opere, 8,* 197.
[68] *On Mechanics*, p. 172; *Opere, 2,* 181.
[69] *Dialogue*, p. 28; *Opere, 7,* 52.
[70] *Dialogue*, p. 203; *Opere, 7,* 229.
[71] *Dialogue*, pp. 207–8; *opere, 7,* 234.
[72] *Discourses, Opere, 8,* 51.
[73] *Discourses, Opere, 8,* 274.
[74] *Loc. cit.*, italics mine.
[75] *Discourses, Opere, 8,* 275.
[76] See T. P. McTighe: 'Galileo's Platonism: A Reconsideration,' in *Galileo, Man of Science* (ed. by E. McMullin), New York 1967, 365–87, and also the Introduction, p. 29.
[77] *Dialogue*, p. 15; *Opere, 7,* 39.
[78] *Discourse, Opere, 8,* 117 seq.
[79] *Opere, 18,* 12–13; translated in S. Drake, 'Galileo's New Science of Motion,' in *Reason, Experiment, and Mysticism, op. cit.*, p. 156.
[80] 'Galileo and Reasoning '*ex suppositione*',' in *Boston Studies in the Philosophy of Science* (PSA 1974), XXXII, 1976, pp. 79–104. See pp. 81, 95.
[81] *Op. cit.*, pp. 99–100.
[82] *Opere, 17,* 90.
[83] *Op. cit.*, p. 92.
[84] *Bodies in Water*, pp. 34 seq., *Opere, 4,* 97 seq.; lodestone and armature, *Dialogue*, p. 407, *Opere, 7,* 433. For a useful survey of the present state of the evidence on the inductive dimension of Galileo's work, see W. L. Wisan, 'The New Science of Motion: A study of Galileo's *De Motu Locali,*' *Arch. Hist. Exact Science* **13** (1974), 103–306; see 'The Role of Experiment,' pp. 120–5.
[85] Letter to Ingoli, 1624, *Opere, 6,* 633; transl. W. R. Shea, *op. cit.*, pp. 148–9. See also *Consideration of the Copernican Hypothesis of 1616, Opere, 5,* 356 seq.
[86] *Op. cit.*, p. 84.
[87] *Dialogue*, p. 145; *Opere, 7,* 171.
[88] S. Drake, 'Galileo's New Science of Motion,' p. 153.
[89] *Op. cit.*, pp. 153–4.
[90] See E. McMullin, 'Structural Explanation,' *American Philosophical Quarterly*, 1978, to appear.

[91] *Discourses, Opere, 8,* 202.
[92] *Loc. cit.*
[93] *Dialogue,* p. 234; *Opere, 7,* 260.
[94] *Dialogue,* p. 481.
[95] This leads Wallace to remark (*op. cit.*, p. 98) that Galileo has a new kind of demonstration since it can work in both directions. But this is peculiar to the context of mechanical laws, and would not work elsewhere. His rendering of the argument in symbolic forms as: 'If q, then if q then p, then p,' is misleading, since this is not an argument-form: it has no conclusion. For it to have a conclusion both q, and (if q then p) would have to be separately asserted. And at this point the apparently "necessary" character of the argument (given it by the local form of a tautology) would vanish.
[96] *Discourses, Opere, 8,* 50.
[97] Letter to Baliani, *Opere, 18,* 13.
[98] In the 'geometrodynamics' of J. A. Wheeler, for example.
[99] *Assayer,* p. 247; *Opere, 6,* 291.
[100] See E. McMullin, 'Structural Explanation,' section 1.
[101] *Assayer,* pp. 310–1; *Opere, 6,* 349.
[102] *Assayer,* p. 312; *Opere, 6,* 351.
[103] *Assayer,* p. 252; *Opere, 6,* 296.
[104] Postil 137 to O. Grassi, *Ratio Ponderum Librae et Simbellae,* 1626, p. 480. Quoted by Shea, *op. cit.*, p. 100.
[105] *Discourses, Opere, 8,* 66.
[106] *Dialogue,* p. 62; *Opere, 7,* 86.
[107] *Loc. cit.*
[108] This is more fully worked out in the *Starry Messenger* than in the *Dialogue.* See Drake translation, pp. 32–40; *Opere, 3,* 63–71.
[109] *Dialogue,* p. 73; *Opere, 7,* 98.
[110] *Dialogue,* p. 76; *Opere, 7,* 101.
[111] *Dialogue,* p. 86; *Opere, 7,* 111.
[112] *Dialogue,* p. 99; *Opere, 7,* 124.
[113] Shea gives a useful account of the methodological issues involved in this controversy in 'Sunspots and Inconstant Heavens,' chapter 3, *op. cit.*
[114] First letter to Welser, Drake, *Discoveries and Opinions of Galileo,* p. 99; *Opere, 5,* 106.
[115] *Op. cit.*, p. 100; *Opere, 5,* 108.
[116] *Op. cit.*, p. 112; *Opere, 5,* 133.
[117] *Op. cit.*, p. 109; *Opere, 5,* 118.
[118] *Dialogue,* p. 54; *Opere, 7,* 79. My italics.
[119] *Loc. cit.*
[120] *Assayer,* p. 331; *Opere, 6,* 368.
[121] *Assayer,* p. 272; *Opere, 6,* 313. He notes here that optics has given him a "necessarily conclusive proof."
[122] *Discourse on the Comets,* Drake translation, p. 50; *Opere, 6,* 88.
[123] *Discourse,* p. 27; *Opere, 6,* 51.
[124] *Discourse,* p. 57; *Opere, 6,* 98.
[125] *Assayer,* p. 260; *Opere, 6,* 303.
[126] *Dialogue,* p. 114; *Opere, 7,* 140.

[127] See the Introduction to *Galileo, Man of Science*, pp. 38–41.
[128] *Dialogue*, p. 118; *Opere*, *7*, 144.
[129] *Dialogue*, p. 122; *Opere*, *7*, 148.
[130] This is discussed more fully in the Introduction to *Galileo, Man of Science*, pp. 33–5.
[131] *Letter to the Grand Duchess*, Drake translation, p. 182; *Opere*, *5*, 317; see also Galileo's letter to P. Dini, *5*, 298 seq.
[132] *Dialogue*, pp. 157–8; *Opere*, *7*, 183–4.
[133] *Against Method*, p. 87.
[134] *Op. cit.*, p. 85. It is not clear to me why showing these notions (e.g. absolute and relative motion) to be inconsistent with one another if taken as parts of the same paradigm, should not be held to put proper pressure on their user to attempt some kind of paradigm revision. Feyerabend rejects my suggestion to this effect in my 'History and Philosophy of Science: A Taxonomy of Their Relations,' *Minn. Studies in Phil. of Science* **5** (1970), 12–67; see pp. 34–41.
[135] As does P. P. Wiener in 'The Tradition behind Galileo's Methodology', *Osiris* **1** (1936), 733–46.

INDEX

[This index departs in some ways from standard analytical indices, but not in ways that will reduce its usefulness. Not all of the papers in the volume could be written easily to conform to the Reidel style sheet. This means that in many cases there is a great proliferation of proper names and titles of books not necessarily required in a simpler style. Thus many proper names have not been indexed, and almost no titles of books and other works. Also, in those cases where there is much reference material, the texts have been left to speak pretty much for themselves.]

Acceleration: and Impetus, 144.
Achinoia: 214.
Albert of Saxony: 87.
Analysis: 217; method of 9, 10, 30; Pappus' use of, 29.
Anamnesis: 251.
Aquinas, T.: 90.
Archimedes: 3, 6, 7, 8, 16, 44, 192, 217, 221, 231; theory of spirals, 40, 41; theory of the lever, 156.
Aristarchus: 219.
Aristotle: 6, 7, 16, 22, 23, 60, 89, 90, 162, 211, 212, 213, 214, 216, 222, 238, 248; his distinction between mathematics and physics, 211, 235; and knowledge of "what" (oti, quia) and of "why" (dioti, propter quid), 215; scientific conclusions as evident principles and as sense dependent, 44, 218; Galileo's rejection of his physics, 27, 186, 218.
Aristotelians: 4, 59, 89, 192, 197, 211, 217.
Atomism: 197; Galileo's early problems with, 198; mathematical, 199, 204; realm of the very small, 241; fire-corpuscles, 241.
Augustine: 249.
Averroës: 90.
Axioms: 7, 226.

Bacon, F.: 5, 47.
Baliani, G. B.: 42, 234, 237.
Beeckman, I.: 139, 140, 142, 143, 144, 147, 149, 157.
Bellarmine, Cardinal: 127; *passim.*
Bible, The: 24, 250; and natural science, 249.
Book of Nature: 236; *passim.*
Boyle, R. 211.: Boyle's Law, 237.
Buonamici, F.: 88, 89, 163.
Buridan, J.: 87, 125, 143.
Burtt, E. A.: 65; *passim.*

Carugo, A.: 164.
Castelli, B.: 30.
Cause: proximate, 214; as true knowledge, 9, 34, 174, 214; Galileo abandons, 162, 223, 237; causal language present in Galileo, 162; Galileo's unconcern for efficient causes, 162; in optics, 165, 166; principles of efficient causality, 170; final causes and the void, 171; and potentiality, 171; Galileo abandons final causes, 193. *See also* Explanation, Geometry, Mathematics, Science, Role of.
Cavalieri, B.: 154.
Clavelin, M.: 54, 223, 237.
Clavius, C.: 93. 95ff., 111, 164.

Cohesion: and the void, 199, 200.
Collegio Romano: 137; *passim*; possible influence on Galileo, 120.
Comets: 122, 245.
Commandinus: 10.
Conceptions of Science: two different in Galileo, 219.
Conceptual Framework: 185; Sellar's account, 185ff.
Conventionalism: Galileo's opposition to, 74ff. *See also* Realism.
Copernicus, N.: 4, 14, 37, 60, 186, 209, 219, 247, 249; system incompatible with the *Bible*, 24.
Corporeal substance: primary and secondary qualities, 64ff. *See also* Qualities.
Crombie, A. C.: 1, 95, 135, 164, 254.

Darwin, C.: 211.
Dating: problem with early Galileo mss., 112. *See also* Style.
De Carcavi: 234, 237.
Deduction: 3, 4, 213; Descartes' methodological reliance on, 155, 222.
Descartes, R.: 28, 60, 66, 88, 222, 224, 241, 248, 252; claims discovery of two laws stated by Galileo, 140; condescending approach to Galileo, 148; Disclaims borrowing from Galileo, 149; instantaneous propagation of light, 152; matter as extension, 152ff.; objection to Galileo's identification of mathematical entities with physical bodies, 154, 155.
Doctores Parisienses: 87, 89, 92, 103, 116; possible source of Galileo's references to, 108ff.; and nominalism, 124.
Drake, S.: 47, 65, 162; *passim*, 237, 252.
Duhem, P.: 75, 88, 89, 112, 124; criticisms of continuity thesis, 87.

Earth's rotation: 248; demonstrated on theological grounds, 250; Galileo's arguments for rejected by Descartes, 139.
Einstein, A.: 75, 76.
Essence: as necessary connection, 214, 222; as intrinsic nature, 166.
Euclid: 28, 29, 30; *passim*, 226.
Experiment: as testing, 223; role in Galileo's science, 11, 12, 224, 226; with lodestones, 33; and free fall, 40ff., 239; inclined plane, 43; methodological importance of, 59, 68. *See also* Explanation, Mathematics, Science, Role of.
Explanation: 217; Hypothetico-deductive, 227; causal or retroductive, 216, 227, 238; inductive, 227; method of analogy, 242; demonstrative, 5, 192, 247; mathematical, 79, 225. *See also* Mathematics, Science, Role of.
Ex suppositione reasoning: 44, 53, 131, 236; hypothetical for Galileo, 234, 237; hypothetical for Bellarmine, 127.

Falling bodies, law of: 13, 126, 141ff., 237, 239; free fall law, 39ff., 79ff., 145.
Favaro, A.: 87, 88, 89, 112, 114, 124.
Fermat, P.: 234.
Feyerabend, P.: 59, 60, 62, 64, 209, 210, 251.

Galen: 122.
Geometry: 10, 187; conventionalism, 75ff.; demonstrative of necessary connections, 174, 189, 221; as non-Aristotelian physical description, 187ff.; paradigm of science, 213; rationale not in syllogistic logic, 28; true as demonstrations of formal causes 174–175. *See also* Mathematics, Science, Role of.
Gilbert, W.: 33, 34; Galileo's discussion of magnetism, 32ff.; 191, 220.
Grassi, O.: 241, 245, 246, 247.
Gravity: 146.
Grümbaum, A.: 76, 77.
Guidobaldo [del Monte]: 10.
Guiducci, M.: 24, 246.

Hales, W.: 116.

Hero: 197.
Huygens, C.: 150.
Hydrostatics: 6ff.

Idealization: 233.
Impediments: 231, 232, 233; as accidents, 22; material hindrances, 8, 73, 74, 75. *See also* Matter.
Imperfection: of matter, 73, 168; 230. *See also* Matter.
Induction: 235; as epagōgē, 213. *See also* Explanation.
Inertia: Descartes' analysis of, 146ff.
Infinity: and uncountable voids, 202.
Ingoli, F,: 226.

Jardine, N.: 253.
Juvenilia: attributed to Gelileo as early works, 88; not early works of Galileo, 123.

Kant, I.: 63, 210, 224.
Kepler, J.: 163; *passim*, 239, 248.
Knowledge: in Aristotle, of the fact and of the reasoned fact, 173, 188; *passim*.
Kuhn, T. S.: 181ff., 188; theories and paradigms, 183; theory refinement and normal science, 184.

Lavoisier, A. L.: 211.
Leibniz, G. W.: 60, 157.
Light, speed of: views of Descartes and Galileo, 151ff.
Locke, J.: 65, 241.
Lokert, G.: 88.
Lucretius: 197.
Lunar surface: resembles that of earth, 242.

Mach, E.: 210.
Mathematicians: 4, 6; and Aristotle's method, 31; as medieval astronomers and theorists of light, 212.
Mathematics: 7; applicable to nature, 45, 70ff.; as demonstration of formal causes, 176; and empirical reasoning, 25ff., 168; and God's understanding, 61, 174, 188; Greek model, 3; proceeds by small and easy steps, 35; no problem for Galileo in applying to the world, 177, 190, 203; and reality, 61. *See also* Cause, Geometry.
Matter: as extension, 152ff.; as atomic, 199, 204; as mathematical in form, 199.
Mechanics: 186, 219, 221, 226, 228; Descartes' criticism of Galileo, 155–156; founded in geometry, 226, 240; as the science of weights, 166.
Medicean Planets: 235.
Mersenne, M.: 139, 140, 143, 145, 147, 148, 149, 155, 156.
Method: resolutive, 9, 29. *See also* Cause, Induction, Geometry, Mathematics, Science, Role of.
Mill, J. S.: 5, 47, 225, 235.
Mixed (middle) sciences: 161, 162ff., 212; Galileo's involvement in these sciences eliminates appeal to efficient causes, 167.
Motion: 6, 19, 217; Galileo's theory in *Two New Sciences*, 37ff.
Murdoch, J.: 128.
Music: significance for 17th-century thinkers, 149.

Naylor, R. H.: 48.
Necessary truths: 225; not acquired by experience, 174; as demonstrative, 174; as results of observation, 22, 23. *See also* Geometry, Mathematics, Science, Role of.
Newton, I.: 46, 88, 143, 182, 238, 239, 241.
Nominalism: 126.

Ockham, W. of: 125.
Optics: 164ff.

Paduan School: 217.
Pappus: 9, 10, 28, 30, 31, 33.
Peirce, C. S.: 227.
Pererius, B.: 93ff.

Plato: 24, 61, 90, 155, 212, 213, 217, 236.
Platonism: 4, 61, 232; reputed to Galileo, 47, 60.
Poincaré, H.: 75, 76.
Pope Urban: 250.
Popkin, R.: 50.
Posterior Analytics: 217, 220.
Premisses: internal simplicity, 229; observational verification, 229; self-warranting as axioms, 226.
Principles: 229; for Galileo and Aristotle no scientific demonstration of, 44.
Proof: 217.
Propaganda: 59, 64; in theory assessment, 210.
Ptolemy: 14, 23, 90; *passim*, 248.

Qualities: primary-secondary distinction, 197; non-subjectivity of secondaries, 65ff.; role in Galileo's methodology, 68ff.

Randall, J.: 253; *passim*.
Realism: in Galileo's conception of science, 74ff.; 129, 191; Galileo's rejection of the instrumentalism of Ptolemy and Osiander, 193. *See also* Conventionalism.
Rocco, A.: 226.
rota Aristotelis: paradox of, 200–202; Galileo's use to solve problem of condensation, 204–205.

St. John Chrysostom: 111.
Sarpi, P.: 11, 12.
Scaliger, J.: 117, 118.
Scheiner, C.: 20, 21, 245.
Science, Role of: as *apodeixis*, 213, 214, 216; as axiomatic, 214; as demonstration, 5, 16, 43, 211, 239, 240, 241, 247; as *epistēmē*, 213, 215, 251; as necessary truth, 174, 219; as study of reasons (dioti), 215, 216; as syllogistic, 214. *See also* Explanation, Geometry, Mathematics.
Sellars, W.: 185, 187.
Settle, T. B.: 48, 135.
Shea, W.: 59, 60, 68.
Statics: 221.
Strong, E. W.: 254.
Style: importance of Galileo's characteristic style in mss. dating, 97ff. *See also* Dating.
Sunspots: 20ff., 245.
Supi, P.: 142.

Telescope: 20; *passim*, 240.
Themo Judaeus: 88.
Theory construction: active role of scientist, 210; axiomatic, 214; inductivistic account, 210.
Tycho Brahe: 25; *passim*, 246, 247, 249.

Universe: age of, 114ff.

Valerio, L.: 13.
Valla, P.: 110.
Vitelleschi, M.: 108ff.

Wallace, W.: 162, 163, 234, 236.
Welser, M.: 20, 21.
Whewell, W.: 210.

Zabarella, J.: 18, 30, 48, 217.

THE UNIVERSITY OF WESTERN ONTARIO SERIES IN PHILOSOPHY OF SCIENCE

A Series of Books on Philosophy of Science, Methodology, and Epistemology published in connection with the University of Western Ontario Philosophy of Science Programme

1. J. Leach, R. Butts, and G. Pearce (eds.), *Science, Decision and Value.* Proceedings of the Fifth University of Western Ontario Philosophy Colloquium, 1969. 1973, vii + 213 pp.
2. C. A. Hooker (ed.), *Contemporary Research in the Foundations and Philosophy of Quantum Theory.* Proceedings of a Conference held at the University of Western Ontario, London, Canada, 1973. xx + 385 pp.
3. J. Bub, *The Interpretation of Quantum Mechanics.* 1974, ix + 155 pp.
4. D. Hockney, W. Harper, and B. Freed (eds.), *Contemporary Research in Philosophical Logic and Linguistic Semantics.* Proceedings of a Conference held at the University of Western Ontario, London, Canada. 1975, vii + 332 pp.
5. C. A. Hooker (ed.), *The Logico-Algebraic Approach to Quantum Mechanics.* 1975, xv + 607 pp.
6. W. L. Harper and C. A. Hooker (eds.), *Foundations of Probability Theory, Statistical Inference, and Statistical Theories of Science,* 3 Volumes. Vol. I: *Foundations and Philosophy of Epistemic Applications of Probability Theory.* 1976, xi + 308 pp. Vol. II: *Foundations and Philosophy of Statistical Inference.* 1976, xi + 455 pp. Vol. III: *Foundations and Philosophy of Statistical Theories in the Physical Sciences.* 1976, xii + 241 pp.

8. J. M. Nicholas (ed.), *Images, Perception, and Knowledge*. Papers deriving from and related to the Philosophy of Science Workshop at Ontario, Canada, May 1974. 1977, ix + 309 pp.
9. R. E. Butts and J. Hintikka (eds.), *Logic, Foundations of Mathematics, and Computability Theory*. Part One of the Proceedings of the Fifth International Congress of Logic, Methodology and Philosophy of Science, London, Ontario, Canada, 1975. 1977, x + 406 pp.
10. R. E. Butts and J. Hintikka (eds.), *Foundational Problems in the Special Sciences*. Part Two of the Proceedings of the Fifth International Congress of Logic, Methodology and Philosophy of Science, London, Ontario, Canada, 1975. 1977, x + 427 pp.
11. R. E. Butts and J. Hintikka (eds.), *Basic Problems in Methodology and Linguistics*. Part Three of the Proceedings of the Fifth International Congress of Logic, Methodology and Philosophy of Science, London, Ontario, Canada, 1975. 1977, x + 321 pp.
12. R. E. Butts and J. Hintikka (eds.), *Historical and Philosophical Dimensions of Logic, Methodology and Philosophy of Science*. Part Four of the Proceedings of the Fifth International Congress of Logic, Methodology and Philosophy of Science, London, Ontario, Canada, 1975. 1977, x + 336 pp.
13. C. A. Hooker (ed.), *Foundations and Applications of Decision Theory*, 2 volumes. Vol. I: *Theoretical Foundations*. 1978, xxiii+442 pp. Vol. II: *Epistemic and Social Applications*. 1978, xxiii+206 pp.
14. R. E. Butts and J. C. Pitt (eds.), *New Perspectives on Galileo*. Papers deriving from and related to a workshop on Galileo held at Virginia Polytechnic Institute and State University, 1975. 1978, xvi+262 pp.